# Lesson *Study:*

*A Handbook of Teacher-Led Instructional Change*

## Catherine C. Lewis

Published by

Research for Better Schools, Inc.
Philadelphia, PA

This material is based in part upon research supported by the National Science Foundation under grants REC 9814967 and RED-9355857. Any opinions, findings, and conclusions or recommendations expressed in this publication are those of the author and do not necessarily reflect the views of the National Science Foundation.

This publication is a product of the Mid-Atlantic Eisenhower Regional Consortium for Mathematics and Science Education at Research for Better Schools under funding from the U.S. Department of Education, Office of Educational Research and Improvement, grant #R319A000009-01. The content does not necessarily reflect the views of the Department or any other agency of the U.S. Government.

# ACKNOWLEDGMENTS

This handbook is made possible by the generosity of more than 100 Japanese educators who have opened up their research lessons to me over the past nine years, and who have taken time from their enormously busy schedules to be interviewed. I would particularly like to acknowledge the contributions made by Kaoru Fujimitsu, Fumio Hiramatsu, Kyoichi Itoh, Mayumi Ito, Yukinobu Okada, Hiroko Oomasa, Kenzou Otaki, and Harumi Sadatoshi, and by the entire faculties of Iwanishi Elementary School, Kitaissha Elementary School, Komae Elementary School Number Seven, Samukaze Elementary School, Sendagaya Elementary School, Takamiminami Elementary School, and Gakugei, Shizuoka, and Tsukuba Affiliated Elementary Schools. I owe a great debt also to the US schools where educators pioneering lesson study have generously shared their work with me: Bret Harte Middle School in Oakland, CA; Martin Luther King Middle School and Oxford School in Berkeley, CA; Paterson School Number Two in Paterson, NJ; Pine Hollow Middle School in Concord, CA; and the entire San Mateo-Foster City, CA, School District.

Professor Shigefumi Nagano of the University of the Air introduced me to the Tokyo-area schools, and has been an important mentor of my work for more than 20 years. In Nagoya and Osaka, Professors Masami Kajita, Eiji Morita, Yasuhiko Nakano, and Katsumi Ninomiya provided helpful contacts with schools and shared their insights about lesson study.

I wish to express my gratitude to the National Science Foundation, the Spencer Foundation Small Grants Program, the Abe Fellowship Program of the Social Science Research Council, and the American Council of Learned Societies, with funds provided by the Japan Foundation's Center for Global Partnership.

The idea for this handbook came from Patsy Wang-Iverson, a researcher and supporter of mathematics and science education whose remarkable expertise spans the territory from big science to fine nuances of elementary mathematics teaching. Patsy has been the starting point for many of the lesson study efforts emerging in the eastern US.

Early on, conversations with Clea Fernandez, Patsy Wang-Iverson, and Makoto Yoshida shaped my thinking about the handbook in many ways. They shared their experiences working on lesson study in the US, and pushed my thinking about lesson study steps and misconceptions.

In thinking about lesson study's adaptation to US settings, I've benefited enormously from the work and personal example of Makoto Yoshida, whose extensive collaboration with educators at Paterson Public School Number Two, Greenwich Japanese School, and several other US sites has uniquely advanced the understanding and practice of lesson study in the US. My thinking about lesson study has also benefited greatly from collaborations with Motoko Akiba, April Cherrington, Shelley Friedkin, David Foster, Nobuaki Hattori, Jackie Hurd, Bill Jackson, Marcia Linn, Lynn Liptak, Maria Magda, Aki Murata, Mary Pat O'Connell, Rebecca Perry, Barbara Scott, Peter Shwartz, Nancy Songer, Akihiko Takahashi, Ineko Tsuchida, Phil Tucker, Hiroko Uchino, Nick Timpone, and Tad Watanabe.

When I first met the members of the Education Department at Mills College about three years ago, it felt a lot like falling in love. The feeling hasn't worn off. I am lucky to work with colleagues who are so wise and generous. They have provided an important sounding board for the ideas of this handbook, and helped me make many connections to the larger world of educational practice and inquiry.

The voices of the three important guys in my life — Andy, Daniel, and Matthew Leavitt — reverberate throughout this handbook and all my lesson study work. They have deftly rescued manuscript figures from computer near-death, willingly rounded up their friends to dub Japanese research lessons into English, and continuously prodded my thinking about science and about learning.

This handbook is dedicated to Sylvia Kendzior (1936-1997), an extraordinary teacher of children and adults whom I was privileged to observe for five years as part of the Child Development Project. Sylvia heard my very first presentation on lesson study in 1993, and I can still remember her enthusiastic and funny response: "People are always telling teachers to reflect. But they never tell us what to reflect *about*. Finally some substance!" Sylvia kept a steady focus on the important issues, and she knew how to build successful groups — groups where people worked hard, felt valued, and willingly took on the hard work of instructional improvement. I can't think of a better touchstone for the challenging work of lesson study.

# ABOUT THE CONTRIBUTORS

**Lynn Liptak** is Principal, Paterson Public School Number Two, in Paterson, New Jersey, and was a member of School Two's "Math Study Group" that pioneered lesson study in the United States. School Two is currently conducting school-wide lesson study in the area of mathematics. She can be reached at lliptak3@aol.com.

**Tad Watanabe** is an Associate Professor of Education at Pennsylvania State University. He previously taught mathematics at Towson University. Between June 2000 and January 2001, Watanabe spent seven months in Japan, observing numerous lesson study meetings. In the US, he participated as a research lesson commentator at Paterson School Number Two, and, together with a teacher from Japan, planned, taught, revised, and re-taught a fourth grade research lesson on the area of rectangles. He can be reached at txw@psu.edu.

**Makoto Yoshida** is president of Global Education Resources (GER), a New Jersey-based, educational consulting firm dedicated to the improvement of elementary and middle school mathematics teaching and learning. GER's current work includes assistance with lesson study implementation, implementation of problem-solving and open-ended approaches, and development of products to support improvement of mathematics instruction. Yoshida's dissertation, an ethnography of lesson study in a Japanese school, formed the basis for *The Teaching Gap*'s chapter on lesson study and will be published in book form by Lawrence Erlbaum Associates. He can be reached at myoshida@globaledresources.com, Web site: www.globaledresources.com.

# TABLE OF CONTENTS

**Chapter**                                                                                          **Page**

1. **What Is Lesson Study?** ....................................................................................... 1
   School-Based Lesson Study at Komae Elementary
      School Number Seven ...................................................................... 2

2. **Why Lesson Study?  Why Now?** ..................................................................... 7
   Brings Educational Goals and Standards to Life in the Classroom ................ 7
   Promotes Data-Based Improvement ..................................................................... 8
   Targets the Many Student Qualities That Influence Learning ........................ 10
   Creates a Demand for Improvement ................................................................... 11
   Values Teachers  ................................................................................................... 12

3. **From New Jersey to California:  Lesson Study Emerges in the US** ............... 15
   Metamorphosis:  From Mathematics Study Group to Lesson Study at
      Paterson Public School Number Two ........................................................ 15
   Teacher-Led, District-Supported Lesson Study in San Mateo, California ...... 19
   Lesson Study in Paterson and San Mateo:  Similarities and Contrasts ........... 20
   Types of Lesson Study ........................................................................................ 20

4. **What Can Teachers Expect from Lesson Study?** .......................................... 27
   Think Carefully about the Goals of a Particular Lesson, Unit,
      and Subject Area ........................................................................................... 28
   Study and Improve the Best Available Lessons ................................................ 29
   Deepen Our Subject-Matter Knowledge ........................................................... 29
   Think Deeply about Our Long-Term Goals for Students ................................. 31
   Collaboratively Plan Lessons ............................................................................. 34
   Carefully Study Student Learning and Behavior ............................................. 34
   Develop Powerful Instructional Knowledge .................................................... 36
   See One's Own Teaching through the Eyes of Colleagues and Students ....... 37
   Summary ............................................................................................................... 38

5. **Time and Scheduling** ....................................................................................... 41
   Lesson Study:  The Basic Sequence of Activities ............................................ 41
   Can Videotape Be Substituted for Live Observation of Research Lessons? ... 43
   Should Groups Meet during or after School? ................................................... 44
   Intervals between Meetings ................................................................................ 48
   Lesson Study:  One More Burden for Teachers? ............................................... 49

**6. Pioneering Lesson Study in Your School: A Step-by-Step Guide** .................51

   Step 1. Form a Lesson Study Group .................51

   Step 2. Focus the Lesson Study.................55

   Step 3. Plan the Research Lesson.................62

   Step 4. Teach and Observe the Research Lesson.................67

   Step 5. Discuss and Analyze the Research Lesson .................70

   Step 6. Reflect on Your Lesson Study and Plan the Next Steps.................71

**7. Supports for Lesson Study** .................75

   Have a Shared, Frugal Curriculum.................75

   Remain Self-Critical .................76

   Remain Open to Outsiders .................78

   Embrace Mistakes .................78

   Don't Worship Originality .................79

   Avoid the Twin Shoals of Happy Talk and Harping .................80

**8. Misconceptions about Lesson Study**.................83

   Misconception 1: Lesson Study Is Lesson Planning.................83

   Misconception 2: Lesson Study Means Writing Lessons from Scratch .........83

   Misconception 3: Lesson Study Means Writing a Rigid "Script" .................84

   Misconception 4: Lesson Study Is Writing the "Perfect" Lesson
      to Be Spread to Others .................84

   Misconception 5: The Research Lesson Is a Demonstration Lesson or
      Expert Lesson .................85

   Misconception 6: Lesson Study Is Basic Research.................86

**9. Next Steps** .................89

**Appendices**.................95

   1. Blackboard Use and Student Note-Taking: Arts Developed through
      Lesson Study.................97

   2. Plan to Guide Learning in Science .................99

   3. Plan to Guide Learning in Mathematics.................107

   4. Plan to Guide Learning in Language Arts.................121

   5. Plan to Guide Learning (Template).................127

   6. Research Map Template.................131

   7. Selected Resources on Lesson Study .................133

# LIST OF FIGURES

**Figure Number**                                                        **Page**

1. Lesson Study Cycle .................................................................................3
2. Teachers' Activities to Improve Instruction.........................................9
3. Contrasting Views of Professional Development ..............................12
4. Types of Lesson Study in Japan ........................................................21
5. Reflection on Lesson Study................................................................27
6. The Role of Outside Specialists in Japanese Lesson Study ..............32
7. Data Collection during Research Lessons: Examples of Focal Questions..........35
8. Qualities of Effective Professional Development ..............................38
9. Lesson Study Schedule.......................................................................42
10. Why Re-teach Lessons? .....................................................................43
11. A Tale of Two Lessons .......................................................................45
12. It's a Matter of Time: Scheduling Lesson Study at Paterson, NJ School Two ...46
13. Lesson Study Steps.............................................................................52
14. Strategies for Building a Lesson Study Group...................................53
15. Choosing a Research Theme (Main Aim) for Lesson Study...............57
16. Komae School Research Map .............................................................59
17. Examples of Four Levels of Lesson Study Goals ..............................61
18. The Three Concentric Circles of the Plan to Guide Learning............63
19. Protocol for Observation and Discussion of a Research Lesson.........69
20. Examples of Student Reflections *(Hansei)* .....................................77
21. US Teacher Faces Pressure to Change...............................................92
22. Japanese Teachers Face Pressure to Change......................................93

# 1

# What Is Lesson Study?

**Professional development that is going to make a difference to students in the classroom must be teacher-driven and student-focused. Lesson study is both of these things.**

- Principal Lynn Liptak, Paterson Public School Number Two[1]

This handbook provides an introduction to lesson study, a Japanese approach to instructional improvement that has recently sparked much interest in the United States (US). As we will see, lesson study is a cycle in which teachers work together to consider their long-term goals for students, bring those goals to life in actual "research lessons," and collaboratively observe, discuss, and refine the lessons.

Since 1993, with research support from the National Science Foundation, I have recorded research lessons in more than 50 schools across Japan and interviewed about 100 Japanese educators. I have also followed lesson study's emergence in a half-dozen US settings, and have closely studied one site, San Mateo-Foster City School District.

Lesson study is a simple idea. If you want to improve instruction, what could be more obvious than collaborating with fellow teachers to plan, observe, and reflect on lessons? While it may be a simple idea, lesson study is a complex process, supported by collaborative goal-setting, careful data collection on student learning, and protocols that enable productive discussion of difficult issues. Hence a handbook is needed.

The first chapter of this handbook provides an overview of lesson study and an example of its use in a Japanese elementary school. Chapter 2 asks "Why lesson study? Why now?" The next section of the book describes the actual practice of lesson study, asking what teachers can expect from lesson study (Chapter 4) and how to schedule, conduct, and support lesson study (Chapters 5, 6 and 7). The final two chapters address common misconceptions about lesson study (Chapter 8) and the future of lesson study in the US (Chapter 9).

My own interest in lesson study began in an odd way. As I sat in Japanese elementary classrooms finishing up a study of Japanese elementary school life,[2] I found myself learning much science, though science had nothing to do with my research focus. Captivated by students' hands-on experiments and their vigorous debates about physical science, I suddenly began to notice levers and pendulums everywhere — can openers, swingsets, my long-handled suitcase. I asked Japanese teachers, who are not science specialists, how they learned to teach in a way that sparked connections to daily life. Over and over again, the answer I heard was "research lessons."

Over the past eight years, I have observed more than 70 Japanese "research lessons," actual classroom lessons that are planned, observed, and discussed by a group of teachers who are trying to bring their educational ideals to life in the classroom.[3] Research lessons are the centerpiece of "lesson study," a teacher-led instructional improvement cycle pictured in Figure 1. In lesson study, teachers work together to:

- Formulate goals for student learning and long-term development.

- Collaboratively plan a "research lesson" designed to bring to life these goals.

- Conduct the lesson, with one team member teaching and others gathering evidence on student learning and development.

- Discuss the evidence gathered during the lesson, using it to improve the lesson, the unit, and instruction more generally.

- Teach the revised lesson in another classroom, if desired, and study and improve it again.[4]

Research lessons occur in many settings in Japan. Think of the various settings for professional development in the US, such as individual schools, district-supported workshops, and professional conferences. Lesson study takes place in all these settings and more, taking on somewhat different characteristics in each. First we examine school-based lesson study, the most basic and widespread type of lesson study in Japan.

## School-Based Lesson Study at Komae Elementary School Number Seven

Komae Elementary School Number Seven, on the outskirts of Tokyo, is a neighborhood public school serving a largely middle-class community. The Komae faculty, who have decided to focus their lesson study on science this year, consider the following two questions in order to choose their "research theme" (main aim) for lesson study:

- Ideally, what qualities do we hope our students will have when they graduate from our school as sixth graders?

- What are the actual qualities of our students now?

Teachers individually think about these questions and jot down their responses, which they later list on the blackboard in two columns: "profile of ideal student" and "profile of actual student." In the "ideal" column teachers list qualities like "love learning," "have deep friendships," "take initiative as learners," and "hold own ideas." Under "actual" they list phrases like "friendly and kind-hearted," "enjoy video games and computers," "some students lack close friendships," and "some students think for themselves but others don't."

Comparing the two lists, teachers discuss the gaps between the "ideal" and "actual" student profiles. One teacher remarks, "I'd like our students to develop their own ideas and perspectives about what they learn, but in fact, many students happily go along with whatever their classmates say, especially if a knowledgeable classmate speaks up." Several teachers mention the impact of video games, computers, and television on their students, remarking that these pursuits infringe on the time children spend playing with each other and playing out-

# Figure 1
# Lesson Study Cycle

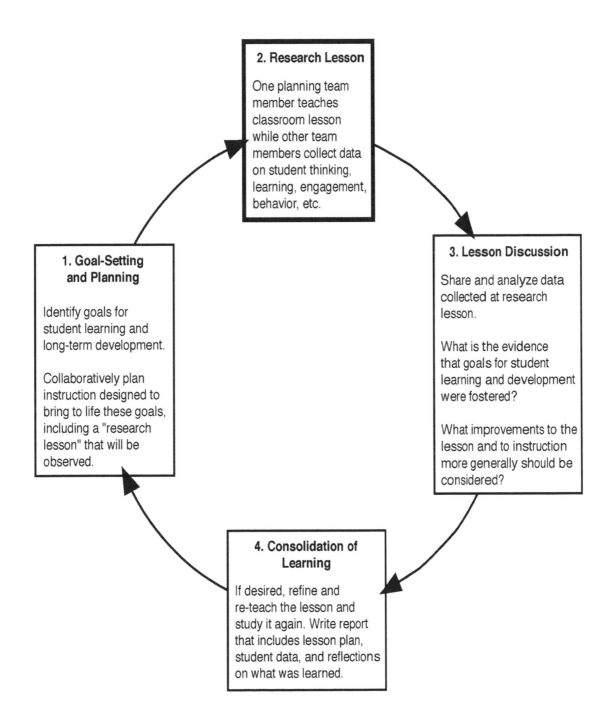

**2. Research Lesson**

One planning team member teaches classroom lesson while other team members collect data on student thinking, learning, engagement, behavior, etc.

**1. Goal-Setting and Planning**

Identify goals for student learning and long-term development.

Collaboratively plan instruction designed to bring to life these goals, including a "research lesson" that will be observed.

**3. Lesson Discussion**

Share and analyze data collected at research lesson.

What is the evidence that goals for student learning and development were fostered?

What improvements to the lesson and to instruction more generally should be considered?

**4. Consolidation of Learning**

If desired, refine and re-teach the lesson and study it again. Write report that includes lesson plan, student data, and reflections on what was learned.

*What Is Lesson Study?*

doors. Teachers comment that electronic entertainment has reduced both children's knowledge of the natural world and the depth of their friendships. From these discussions, Komae teachers develop their research theme: "For students to value friendship, develop their own perspectives and ways of thinking, and enjoy science."

Over the next few months, Komae teachers meet in three separate groups, with lower-, middle-, and upper-grade teachers each forming a group. Each group picks a science unit taught at their level and plans a research lesson within it to be observed, studied, and discussed by the entire Komae faculty. In planning the unit and research lesson, they seek to bring to life both the long-term goals for children expressed in their research theme (friendship, enjoyment of science, and development of one's own views) and the subject-matter goals for science and for the specific topic (e.g., levers) laid out in the national *Course of Study*. One research lesson, from the grade five and six group, is highlighted on the video *Can You Lift 100 Kilograms?*[5] and its instructional plan is provided in Appendix 2.

As the fifth- and sixth-grade teachers plan the research lesson, they theorize that students "develop their own perspectives and ways of thinking when they face a compelling problem, and personally try to solve it." So the teachers decide to begin the levers unit "by challenging students to lift something really heavy, that can't be lifted with their arms, so they can see the real power of a lever." Teachers plan a lesson that challenges students to lift a 100-kilogram (220-pound) sack of sand. The small desktop balances often used to teach about levers will be introduced only in the second part of the levers unit, when students investigate how to balance a lever.

In another departure from their past teaching of levers, the teachers decide not to provide poles and fulcrums or to suggest that students try levers. Instead they simply pose the challenge of moving a 220-pound sack, expecting that students will devise several methods (such as a pulley, or pushing the sack onto a wheeled cart) that, by contrast, point up the power of the lever. Another innovation, designed to promote students' individual thinking, is to have students first work individually to devise plans to lift the sack, and only later work in groups. Finally, rather than working within their usual small groups,[6] the students will work with other students who have devised similar plans for solving the problem; this modification is designed to deepen students' thinking about the solution method they devised.

The research lesson planned by the grade five-six team and taught by one team member is observed by the whole faculty of Komae School. Each faculty member is assigned to observe and record what goes on within a particular group of students. With enormous enthusiasm but varying degrees of success, the fifth graders struggle to lift the 220-pound sack using the methods they have designed, including pulleys, ramps, and levers. The observing teachers take detailed narrative notes on the discussion and activity of each small group, noting how students' ideas change, whether and how each student in a group participates, and whether students share their ideas across groups. At the faculty colloquium following the lesson, teachers use these data to reflect on the strengths and shortcomings of the lesson with respect to helping students understand levers and with respect to the research theme of helping students "value friendship, develop their own perspectives and ways of thinking, and enjoy science."[7] They recognize a number of strengths of the lesson and some areas that need further attention. For example, their data suggest that even the five students who are normally very quiet spoke up and participated in the group activity, perhaps because the lesson design required every student to lift the sack. On the other hand, teachers observe that not all groups learned from the solutions of other groups, and they conclude that more systematic opportunities for exchange of information among groups would improve the lesson in the future. These notes are written up, and, along with the lesson plan, photographs of the lesson, and examples of student work, form a report on the research lesson that is available in the school office.

At Komae School Number Seven, research lessons are held about six times during the school-year, presented in turn by the lower-, middle-, and upper-grade teachers. Like their counterparts at elementary schools across Japan, Komae's teachers have regular opportunities to see, discuss, and refine instruction designed to bring to life their shared goals for students.

Lesson study is the major form of professional development chosen by Japanese teachers. In the lesson study cycle, teachers collaborate to formulate goals for student learning and development, plan instruction designed to foster these goals, and observe and discuss selected "research lessons." As the next chapter explores, lesson study also provides a powerful mechanism for system-wide improvement of education.

# Chapter 1 Notes

[1] Research for Better Schools (2000, Fall). Against the odds, America's lesson study laboratory emerges. *RBS Currents*, 4:1, 10. (http://www.rbs.org.)

[2] Lewis, C. (1995). *Educating hearts and minds: Reflections on Japanese preschool and elementary education*. New York: Cambridge University Press.

[3] In addition, I interviewed about 100 Japanese classroom teachers and principals; all interviews were conducted in Japanese, and translations are my own unless otherwise noted.

[4] Re-teaching the research lesson is optional, but highly recommended by Makoto Yoshida, a pioneer of lesson study in US schools.

[5] *Can you lift 100 kilograms?* (2000). Video. 18 min. Available from lessonresearch.net. Highlights three parts of the lesson study cycle in Japan, showing Japanese teachers engaged in planning, conducting, and discussing a fifth grade research lesson on levers.

[6] Japanese students often work in mixed-ability, family-like small groups *(han)* which may stay together for months, typically for many different activities of the school day; see Lewis, 1995, for a discussion of the role of these groups in promoting students' learning and attachment to school. However, for this lesson, teachers chose to form new groups made up of students who had devised similar strategies for lifting the sack.

[7] *Can you lift 100 kilograms?* loc. cit.

# 2

# Why Lesson Study?  Why Now?

**Improving something as complex and culturally embedded as teaching requires the efforts of all the players, including students, parents, and politicians.  But teachers must be the primary driving force behind change.  They are the best positioned to understand the problems that students face and to generate possible solutions.**

*- James Stigler and James Hiebert, The Teaching Gap*[1]

At a time when so many school districts are already suffering from reform overload, why has lesson study attracted so much attention?  In *The Teaching Gap*, James Stigler and James Hiebert argue that lesson study supplies a key missing element in reform: an effective way to improve teaching and learning through development of a shared professional knowledge base on teaching.[2]

While most of this handbook is addressed to teachers, this chapter focuses on an issue that is also of keen interest to administrators: lesson study's role in systemic change.  In the US, lesson study is best known as a method to improve classroom lessons.  But in Japan, lesson study contributes not just to teachers' professional knowledge (the topic of Chapter 4), but also to system improvement more broadly.  This chapter highlights the pathways by which lesson study:

- Brings educational goals and standards to life in the classroom.

- Promotes data-based improvement.

- Targets many student qualities that influence learning.

- Creates grassroots demand for instructional improvement.

- Values teachers.

## Brings Educational Goals and Standards to Life in the Classroom

The US now has many standards designed to improve classroom instruction.  How are these standards best brought to life in classrooms?  Top-down mandates and high-stakes assessment have well-known disadvantages, and many common forms of professional development appear to have little impact on instruction.[3]

Lesson study provides a collaborative process for teachers to make sense of educational goals and standards and to bring them to life in the classroom. For example, the Japanese teachers featured in the video *Secret of Trapezes*[4] deliberately designed a lesson in which students were not instructed to control variables in their experiments, and many students failed to do so; students were then helped to discover that the outcomes of the uncontrolled experiments could not be trusted. These teachers had considered two different Japanese national standards — learning scientific habits of mind and learning about the factors that affect pendulum cycle — and tried to design a unit that honored both, by helping students experience the importance of controlling variables as well as acquire particular knowledge about pendulums.

The Japanese do not devote much paper to goals and standards; the entire national *Course of Study* for elementary schools is contained in a small, slim volume (100 half-letter size pages) with similarly-sized volumes for each subject area.[5] However, Japanese educators devote much time and energy to lesson study, through which they bring their frugal national goals and standards to life in the classroom. In contrast, in the US, we tend to spend a great deal of time writing standards and goals and relatively little time studying and refining the classroom lessons designed to bring these ideas to life. Lesson study researcher Clea Fernandez comments on US reform:

> *When reform ideas fail to move from rhetoric to action, we often interpret this as a failure to communicate the ideas clearly, and we revise and improve the documents. It is as if we feel that, if we can find just the right set of words and examples telling teachers what to do in their classrooms, they will act accordingly.*[6]

Figure 2 schematically illustrates the use of instructional improvement time in the US and Japan. Many factors conspire to keep US teachers in the top layer of the triangle, where they spend their time articulating what will be taught at each grade level, finding curricula, trying to align curricula with state or district standards, and writing lessons to fill the resulting holes. The triangle of US instructional improvement thus stands precariously on its tip, without adequate time devoted to observation, discussion, and improvement of actual classroom lessons. In contrast, Japanese instructional improvement rests on a base of observation, discussion, and refinement of classroom lessons. Lesson study is a way to shift emphasis from the top layer of Figure 2 to the base, so that our instructional improvement efforts emphasize lesson observation and improvement.

## Promotes Data-Based Improvement

We are often told that school reform should be guided by data. Unfortunately, the data are often limited to standardized tests that measure a very narrow slice of academic performance. In contrast, during research lessons teachers carefully observe students and collect data to answer questions like the following:

- How did students' knowledge and understanding of the topic change over the course of the lesson and unit?

# Figure 2
## Teachers' Activities to Improve Instruction

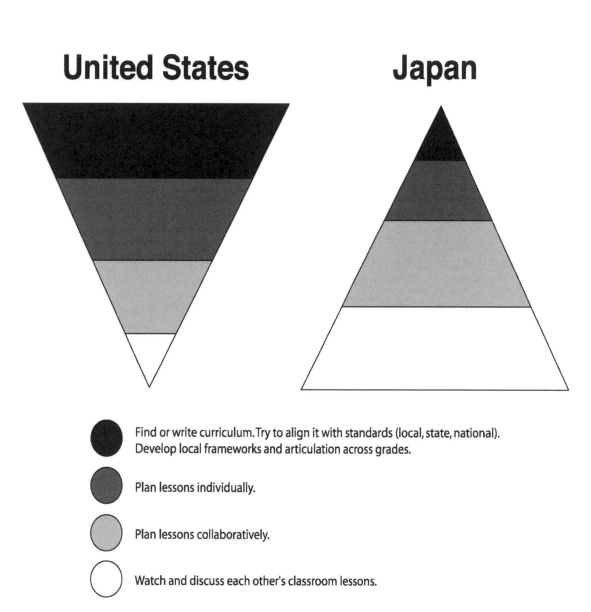

**United States**  **Japan**

Find or write curriculum. Try to align it with standards (local, state, national). Develop local frameworks and articulation across grades.

Plan lessons individually.

Plan lessons collaboratively.

Watch and discuss each other's classroom lessons.

- Are students really interested in this topic, or are they just going through the motions?

- Do students possess the basic personal qualities needed for learning? For example, are they well-organized, responsible and able to listen and respond to one another's ideas?

In a process very similar to the "quality circles" credited for the ascent of the Japanese auto industry, teachers analyze these data and use them as a basis to design changes in instruction, classroom procedures, and classroom climate. While lesson study scrutinizes students' academic learning, it also scrutinizes motivation and social climate, factors likely to contribute heavily to students' long-term academic success.[7]

While tests and student work may offer information on *what* to improve, lesson study also sheds light on *how* to improve. For example, observers might notice that the manipulative used to teach a particular concept is misunderstood by students and suggest a better one. In contrast to standardized tests, feedback from lesson study is immediate, specific to the school's curriculum and goals, and based on actual observation of the lesson. It comes from colleagues likely to have intimate knowledge of the students and their context, the people "best positioned to understand the problems that students face and to generate possible solutions."[8]

## Targets the Many Student Qualities That Influence Learning

Think of a classroom you know. Imagine what would happen if every child arrived at the beginning of the year well-organized, eager to uphold class rules, able to get along well with others, and enthusiastic about learning. What difference would it make in what you could teach? The habits of mind and heart that are fundamental to success in school — such as persistence, cooperation, responsibility, and willingness to work hard — develop over many years, in many classrooms. As a Japanese elementary teacher explained, teachers cannot greatly improve children's lives except by working together as a whole faculty to provide a coherent, consistent environment for children's development of these qualities. What's the use if children learn to "think like scientists" in one classroom, only to have those qualities devalued by next year's teacher?

Two American science education researchers who visited Japanese elementary schools with me were less surprised by the science lessons (which, they said, shared a great deal with exemplary US science programs) than by the basic habits and characteristics of the students.[9] For example, students listened and responded to each other's ideas during discussions, responsibly handled dangerous and fragile materials, took careful notes, worked easily with group-mates, and cleaned up the inevitable broken beakers and water spills without attention from the teacher. Although it is tempting to attribute these qualities to family influences, they are all qualities that Japanese teachers systematically address during lesson study, as part of long-term goals for student responsibility, persistence, and friendship. Lesson study does not just target academic development. It targets the many personal qualities that contribute to student motivation and learning, and reshapes many elements of school life in order to promote those qualities.

The long-term focus of lesson study may help prevent pendulum swings between the social and academic goals of education.[10] When teachers look at instruction simultaneously through the lenses of promoting friendships and academic learning, it is likely they will attend to both, rather than teaching in ways that inadvertently undermine one or the other. The history of US education has been plagued by pendulum swings between "self-esteem," on the one hand, and academic rigor on the other, a cycle that can be escaped only if schools learn to promote academic development while also meeting students' fundamental human needs for belonging and contribution.[11] Long-term goals that emphasize both social and academic development may help guard against the "quick fixes" that boost test performance while undermining students' subsequent motivation to learn and their experience of school as a supportive environment.

## Creates Demand for Improvement

Lesson study also ups the ante on what's good teaching. No one requires Japanese teachers to adopt the research lessons they see. But good new approaches tend to spread quickly. During a research lesson, if you see students gasp in amazement when they measure circles of various sizes and discover that the circumference is always about three times as long as the diameter, you will want to give your own students the thrill of discovering pi in this way.

Through live research lessons, written reports, videos, and sharing of experiences with colleagues, lesson study spreads thoughtfully-designed lessons on a wide range of topics, creating a system that learns.[12] The cumulative effect of research lessons that ripple through the system is locally-initiated, locally-managed system-wide reform.[13] For example, inspired and informed by the research lessons of a group of nationally active teachers in the 1970s, teachers all over Japan began to emphasize problem solving in mathematics, gradually leading to a widespread shift toward "teaching for understanding" in elementary mathematics over the past three decades.[14]

It has been said that American education suffers not from a shortage of good programs, but from a lack of demand for them.[15] Ironically, the lively Japanese science lessons that first interested me in lesson study were based in part on exemplary US science lessons observed by Japanese educators visiting the US. These approaches spread widely in Japan through the system of research lessons.

Research lessons create a natural grassroots demand for improvement of teaching. A Japanese teacher recalls how, early in her career, she burst out into tears after seeing a wonderful research lesson taught by her fellow first-grade teacher:

> *I felt so sorry for my own students. I thought their lives*
> *would have been so much better if they'd been in the other*
> *teacher's class. You realize you have had a big impact on*

*your students. You see how authoritarian teachers have very quiet classes. Teachers who value students' ideas have very active classes. You see how teachers are creating a class, not just teaching a lesson. The teacher's way of speaking and the teacher's way of getting angry are all passed on to the students.[16]*

Research lessons provide a chance to study the learning and engagement of students in other classrooms, and to inform and fuel one's own improvement as a classroom teacher.

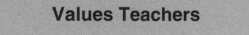

## Values Teachers

The most noble, magnificent educational visions are just spots of ink on paper until a teacher somewhere brings them to life in a classroom. Lesson study recognizes the central importance and difficulty of teaching — of actually bringing to life standards, frameworks, and "best practices" in the classroom. Lesson study invests time and resources in planning, studying, and refining what actually happens in classrooms. It is a system of research and development in which teachers advance theory and practice through the careful study of their own classrooms, constantly testing and improving on "best practices." Lynn Liptak, one of the first US principals to implement lesson study, contrasts traditional professional development and lesson study in Figure 3. As it shows, lesson study places teachers in an active role as researchers.

**Figure 3**
**Contrasting Views of Professional Development***

| Traditional | Lesson Study |
| --- | --- |
| • Begins with answer | • Begins with question |
| • Driven by outside "expert" | • Driven by participants |
| • Communication flow: trainer → teachers | • Communication flow: among teachers |
| • Hierarchical relations between trainer & learners | • Reciprocal relations among learners |
| • Research informs practice | • Practice *is* research |

*by Lynn Liptak

The Japanese classroom teachers most actively involved in lesson study regularly attract thousands of educators to their research lessons, and write widely-read books about their lessons. Lesson study allows Japanese teachers to enjoy the satisfactions of classroom research and to influence education broadly through their research lessons, while keeping their feet firmly planted in the realities of classroom life. As one US teacher said about lesson study, "One of the things that I really love about it is that it puts a professional part back in teaching that we have to battle for all the time... being able to say 'This is like a science, and we can figure these things out and get better at them.'"[17] The next chapter visits two sites that have pioneered lesson study's adaptation to the US.

# Chapter 2 Notes

[1] Stigler, J.W. & Hiebert, J. (1999). *The teaching gap.* New York: Free Press, 135.

[2] Ibid., 131.

[3] Cohen, D.K. & Hill, H. (2002). *Learning policy.* New Haven: Yale University Press; Loucks-Horsley, S., Hewson, P.W., Love, N., & Stiles, K.E. (1998). *Designing professional development for teachers of science and mathematics.* Thousand Oaks, CA: Corwin Press.

[4] *The Secret of Trapezes* (1999). Video. 16 min. Available from lessonresearch.net. See also Lewis, C., & Tsuchida, I. (1998, Winter). A lesson is like a swiftly flowing river: Research lessons and the improvement of Japanese education. *American Educator*, 14-17 & 50-52.

[5] Japanese Ministry of Education (1998). *Course of study for elementary schools (Shougakkou gakushu shidou youryou).* Tokyo: Government of Japan Printing Office. This is a half-letter size booklet of 100 pages that lays out the minimum hours, goals, and content for all 12 required content areas (Japanese, social studies, mathematics, science, life environment studies, music, art, homemaking, physical education, ethics, special activities, and interdisciplinary studies). It is available at http://www.mext.go.jp/a_menu/shotou/youryou/index.htm.

[6] Clea Fernandez, Personal Communication, May 8, 2001; Clea Fernandez is a founder of the Lesson Study Research Group at Teachers College Columbia University (http://www.tc.columbia.edu/lessonstudy/).

[7] Solomon, D., Battistich, D., Watson, M., Schaps, E., & Lewis, C. (2000). A six-district study of educational change: Direct and mediated effects of the child development project. *Social Psychology of Education,* 4, 3-51.

[8] Stigler & Hiebert, op. cit., 135.

[9] Linn, M., Lewis, C., Tsuchida, I., & Songer, N. (2000). Science lessons and beyond: Why do US and Japanese students diverge in science achievement? *Educational Researcher,* 29, 4-14.

[10] Lewis, C. & Tsuchida, I. (1997). Planned educational change in Japan: The shift to student-centered elementary science. *Journal of Educational Policy,* 12:5, 313-331.

[11] Research suggests that schools can promote both social and academic development effectively if they use intrinsic motivation, rather than competition, to foster academic development. See, for example: Lewis, C., Schaps, E., & Watson, M. (1999). Recapturing education's full mission: Educating for social, ethical, and intellectual development. In C. Reigeluth (Ed.), *Instructional-design theories and models*: *A new paradigm of instructional theory.* Mahwah, NJ: Lawrence Erlbaum Associates, 511-539; Solomon, Battistich, Watson, Schaps, & Lewis, loc. cit.; Lewis, C., Schaps, E., & Watson, M. (1995, March). Beyond the pendulum: Creating caring and challenging schools. *Phi Delta Kappan*, 76, 547-554.

[12] See Stigler & Hiebert, op. cit.

[13] Lewis & Tsuchida (1999), loc. cit.

[14] Takahashi, A. (2000). Current trends and issues in lesson study in Japan and the United States. *Journal of Japan Society of Mathematical Education*, 82: 12, 15-21. (In Japanese).

[15] Elmore, R. (1999-2000, Winter). Building a new structure for school leadership. *American Educator*, 23:4, 12.

[16] Lewis & Tsuchida (1998, Winter), op. cit., 17.

[17] Jacqueline Hurd, Interview, May 10, 2001.

# 3

# From New Jersey to California: Lesson Study Emerges in the US

> Lesson study is teacher-directed, teacher-driven.... It is really teacher-oriented. Most other professional development is like a seminar. You sit there and you listen. You may do a little bit of hands-on stuff, but usually they are just feeding you information. Here, we are seeking our own information. We are doing our own research.
>
> - Heather Crawford, Teacher[1]

Although I began studying research lessons in 1993, there was little interest in the practice until the Third International Mathematics and Science Study (TIMSS, 1995) documented dramatically different styles of mathematics instruction in US and Japanese classrooms and high student achievement in Japan.[2] *The Teaching Gap* attributed Japan's effective lessons to lesson study, and brought Makoto Yoshida's ethnography of lesson study to a broad public audience.[3] Since then, dozens of lesson study efforts have sprung up across the US.[4] This chapter briefly examines two US lesson study efforts: one in an urban public school in Paterson, New Jersey, and the other in a largely suburban school district near San Francisco, California.

## Metamorphosis: From Mathematics Study Group to Lesson Study at Paterson Public School Number Two[5]

Americans frequently ask whether lesson study, a product of Japan's relatively homogeneous schools, can work in America's more diverse schools. Paterson Public School Number Two in Paterson, New Jersey provides an excellent setting to explore this question, for it faces economic and social challenges common to America's urban schools: 95 percent of the school's 720 pre-K–8 students qualify for free lunch, 99 percent are from minority groups, and over half enter school speaking limited or no English.[6] The school's transience rate is 42 percent, and, as Principal Lynn Liptak pointed out at the school's lesson study open house in the spring of 2000, this pattern means that only three of the school's seventh graders had been attending the school since kindergarten.[7]

The roots of lesson study at Paterson School Number Two stretch back to the spring of 1997, when the principal and eighth grade teachers together attended a district-sponsored workshop and viewed the videotapes of US and Japanese mathematics lessons from TIMSS.[8] Bill Jackson, a School Number Two teacher, recalls the workshop:

*Many of the teachers reacted defensively to the videotaped lessons, seeing TIMSS as yet another indictment of American teachers. But others, including myself, were intrigued by the tapes and interested in learning more. After watching the videotape of the American lesson, I realized that although I had always attempted to incorporate new and interesting teaching methods in my classes, my teaching generally resembled that of the American teachers and was very different from the method used by the Japanese teachers.[9]*

Over the ensuing months, Jackson began experimenting with the pattern of instruction he had seen in the Japanese lesson. He would pose an interesting word problem, have students work on the problem and present and discuss their various solution methods, and draw out the salient mathematical points from their responses. He recalls, "I was excited to see my students engaged in learning and discussing mathematics and surprised to discover that my students could construct mathematical ideas on their own without me explaining the procedure first."[10]

Jackson and fellow teacher Beverly Piekema spent the summer of 1997 developing Japanese-style mathematics lessons aligned with the New Jersey Core Curriculum Content Standards; they then taught the lessons to their school's eighth graders during the 1997-98 school year.[11] Jackson recalls:

*Although these lessons began initially as imitation, they began to evolve into something much deeper. As we taught the lessons, we made notes about their strengths and weaknesses. We also videotaped and analyzed several lessons to begin to look critically at our own teaching. We met with the principal regularly to discuss the progress of the lessons and also participated with other mathematics teachers across grade levels in a mathematics study group organized by the principal. In this study group, teachers met weekly to discuss current research, visit and observe other classes and schools, plan lessons cooperatively, observe one another's teaching, and offer suggestions for mutual improvement.[12]*

Thus was born in October 1997 the Mathematics Study Group, a group of ten volunteer teachers in grades one through eight and Principal Liptak. The group met for one 80-minute session per week during school hours. Liptak recalls that, unlike "grade level meetings [where] we spent too much time on issues that were not directly focused on classroom instruction..., the Mathematics Study Group became something special, where teaching and learning were truly central."[13]

The Mathematics Study Group continued to meet throughout the 1997-98 school year, as teachers tried out, revised, and polished the lessons they had developed, keeping notes on each lesson. Liptak remembers that "the group was particularly enthralled with a videotape of one of Bill Jackson's new lessons, which showed our own eighth grade students now actively involved in thinking deeply about mathematics."[14] In January 1998, Patsy Wang-

Iverson of Research for Better Schools began to participate in the Mathematics Study Group. Both a professional scientist and a technical assistance provider for mathematics and science instruction, Wang-Iverson observed and provided feedback on mathematics lessons.

A teacher recalls the development of the Mathematics Study Group: "We had many opportunities to observe others as well as ourselves. Co-planning was necessary... I did not feel I was alone during this *metamorphosis*."[15] An eighth-grade student at School Two, interviewed by Wang-Iverson in spring 1998, summed up the change in the mathematics instruction this way: "Last year the book did the thinking. This year we do the thinking."[16] James Hiebert, co-author of *The Teaching Gap*, commented on a videotape of a spring 1998 lesson by School Two teacher Bill Jackson:

> *Perhaps the strongest impression is of students engaged in thinking and reasoning mathematically, a critical aspect of learning mathematics well but an aspect that is surprisingly difficult to implement. You have established a culture in your classroom that supports students in really doing mathematics, and many of those who view the tape (including myself) would be eager to know how you accomplished this.*[17]

Wang-Iverson introduced the Study Group to Japanese research lessons through the article "A Lesson Is Like a Swiftly Flowing River"[18] and the tape *Secret of Trapezes*.[19] The Mathematics Study Group sought help from lesson study researchers Clea Fernandez and Makoto Yoshida, who agreed to design and support an intensive partnership between School Two's Mathematics Study Group and teachers from the Greenwich Japanese School.[20] This partnership has, in Liptak's words, "proven invaluable in helping us learn about conducting lesson study as well as providing examples of what powerful mathematics teaching looks like."[21] From January to March 2000, Japanese partner teachers from the Greenwich (Connecticut) Japanese School[22] spent one full day each week at Paterson School Two, where they worked with the Mathematics Study Group (now with 16 voluntary participants and a two-hour weekly meeting) to prepare, observe, and refine research lessons. Choosing as their lesson study goal *to foster students' problem solving and responsibility for learning in mathematics*, the teachers formed four working groups by grade level (K-1, 2-3, 5-6, and 7-8) and each group worked on a research lesson, which they taught, revised, and then taught again publicly in February 2000.

On February 28, 2000, I attended these first-ever US public research lessons. One lesson introduced 19 second graders to multiplication with the following problem:

> *I bought five Kit-Kat bars. There are four chocolate sticks in each bar. How many chocolate sticks in all? Do we have enough for every student in the class to have a piece?*

As students worked to represent and solve the problem at their desks, observers circulated the classroom, taking notes on the strategies students used. Later, the teacher asked several students to share their solutions on the blackboard; each of these students had used a different approach to solve the problem (such as skip-counting by fours, drawing five groups of four, or adding fours and eights to represent one or two bars). These students explained the thinking behind their solutions; for example, one student explained why skip-counting by fours was similar to adding groups of four. After the class discussed each of the approaches on the board, students were given a second multiplication problem to work on individually. This time, observers circulating among the desks could see whether the lesson had helped students progress from a simpler strategy (such as counting by ones) to a more multiplication-like strategy, such as adding groups.

Watching this first-ever public research lesson in a US school, I was struck by how much is communicated during a research lesson. Most obvious, of course, was the lesson itself, a carefully designed lesson that offered insights into multiplication and into the qualities of a motivating, effective lesson. A written lesson plan distributed to all observers summarized the lesson and provided lesson handouts and suggestions about the aspects of student thinking to notice during each phase of the lesson.

The research lesson also communicated that lessons can be progressively improved through work with colleagues. The four teachers who had planned the lesson described how they had revised the lesson's visual aids after teaching the lesson, eventually settling on large laminated blow-ups of the fronts of four Kit-Kat bars and the inside of one. They had redesigned the visual aids in this way so that students would be clear on the numbers for the mathematics problem (five candy bars with four pieces in each) but would not be encouraged to solve the problem simply by counting pieces, as they might if the contents of all five bars were visible. This redesign modeled yet another hallmark of lesson study: anticipating students' reactions to a lesson and planning for them.

The live research lesson enabled observers to study firsthand students' work, engagement, and treatment of each other, and thus to see a lesson from the student point of view. My own ideas about the mathematical capabilities of second-graders expanded as I heard their sophisticated discussions of the similarities between counting by fours and adding 4+4+4.

Finally, I was struck by another quality of the research lesson, namely the power of real, live students, deeply engaged in learning mathematics, to renew and inspire adults. For many of us, the most astonishing moment of the lesson came when the teacher suggested that the class take a break and eat the Kit-Kat bars before solving the second problem. Students, however, were so eager to solve the second multiplication problem that they elected to postpone eating the candy bars. Several observers gasped audibly. A professor standing nearby whispered in my ear, "When was the last time you saw *that* happen?"

Since those first public research lessons in early 2000, lesson study has steadily grown at Paterson Public School Number Two and become central to the work of the Mathematics Study Group, as Principal Lynn Liptak explains:

> *We are finding that lesson study is a powerful form of practical inquiry. It has served to focus us more precisely on the lesson as the unit of change. Developing, testing, discussing, and revising our lessons have become our action research.... Teaching, conducted collaboratively and reflectively, is research.*[23]

Although it is too early to assess the effects of the Mathematics Study Group's work, achievement test data and enrollment of School Two graduates in algebra are certainly consistent with improvement of instruction.[24]  At the very minimum, School Two provides evidence that US teachers in a diverse, urban school can bring lesson study to life and find it useful.

## Teacher-Led, District-Supported Lesson Study in San Mateo, California

In the San Mateo-Foster City School District (SM-FCSD), near San Francisco International Airport, lesson study began in a different way.  Jackie Hurd, a third-grade teacher who spends part-time out of her classroom as a district mathematics coach, experienced lesson study at an international mathematics conference in Japan during the summer of 2000.  Once back in San Mateo, she shared information about lesson study with her three fellow mathematics coaches, a group that had been working together for several years to support improvement of mathematics instruction.  Together they approached the district to request release time and stipends for lesson study volunteers.  The district agreed, drawing on federal and private grant funds for mathematics improvement. The mathematics coaches sent out a one-page description of lesson study to all district teachers, inviting interested volunteers to a meeting to learn about lesson study.  By the fall of the 2000-2001 school year, seven mathematics lesson study teams, including 28 teachers representing eight schools, had been formed.  All groups planned, conducted, and discussed research lessons, and interviews at the end of the school year indicated that teachers found the practice useful.   One district elementary school voted to practice lesson study school-wide and several educators planned to expand lesson study to language arts the next year.

While the Paterson teachers worked with experienced Japanese colleagues and lesson study researchers, the San Mateo mathematics coaches made clear to their colleagues from the start that "we're not experts — we're all learning together how to do lesson study."[25]  Drawing on videotapes,[26] protocols from the Lesson Study Research Group,[27] a visit to research lessons at the Greenwich Japanese School, and advice from a number of researchers, the mathematics coaches both supported and participated as members in the lesson study groups.  At the end of the first year, they also led a two-week summer workshop co-designed with the lesson study team at Mills College and with mathematics specialists.  The workshop's first week was devoted to the study of geometry and the second week to conducting research lessons, with twenty local teachers and six Japanese educators planning, conducting, and discussing research lessons in two year-round elementary schools in the district.  This opportunity to conduct mathematics lesson study with Japanese colleagues further sparked lesson study's development in San Mateo, and by the fall of 2001, 58 teachers had joined lesson study, which had expanded to include language arts as well as mathematics.

## Lesson Study in Paterson and San Mateo: Similarities and Contrasts

The emerging lesson study efforts at Paterson Public School Number Two and SM-FCSD differ in many ways. Paterson's effort is school-based, while the SM-FCSD effort is hosted by district mathematics coaches, and includes both cross-school and school-based groups. Paterson teachers meet during the school day on salaried time while SM-FCSD teachers meet mainly after school and receive a stipend. While the principal was a key player in initiating School Two's effort, SM-FCSD's effort was initiated by classroom teachers and mathematics coaches. But in other important ways, the efforts are similar. In both places, groups already engaged in collaborative work to improve mathematics instruction took up lesson study. Classroom teachers are in the leadership of both efforts, and both have substantial grassroots support from teachers. At the same time, both efforts draw strategically on outside specialists in mathematics instruction and lesson study.

These two lesson study efforts and others that are emerging across the US provide enormously important resources in several ways. They are reservoirs of practice where US educators can see lesson study in action, an experience different from reading about it or seeing it on videotape. These emerging lesson study sites are also laboratories in which US teachers are actively adapting lesson study to US circumstances in response to the challenges and supports encountered in each setting.

A Japanese principal reminds us that building lesson study in the US is long-term work. As he worked with Paterson School Two to build lesson study, he judged their progress as excellent, and predicted that, in a decade, the school would be in a position to teach other US schools about lesson study. Paterson teacher Nick Timpone underlined the idea of lesson study as a long-term process:

> *Understand that lesson study is not a quick fix. It is not a fad diet that you use for six months, give up, and then go back to eating the things that made you fat in the first place. It is a teaching lifestyle change. You need to buy into it for the long haul. It will pay off.*[28]

## Types of Lesson Study

Lesson study takes many different forms in Japan, as shown in Figure 4. The most basic form of lesson study is school-based. Teachers at virtually every Japanese elementary school take part in it. As one Japanese teacher said, "Why do we do research lessons? I don't think there are any laws [requiring it]. But if we didn't do research lessons, we wouldn't be teachers."[29] Another Japanese teacher reflects:

**Figure 4**
**Types of Lesson Study in Japan**

| |
|---|
| **Sponsor: A public or private school** |
| **Examples:** The school faculty agrees on a shared, school-wide goal, and brings it to life in "research lessons" planned by grade-level groups. These lessons (3–6 times/year) are observed and discussed by the whole faculty. |
| **Sponsor: Private, voluntary teacher circles** |
| **Examples:** Teachers meet regularly to discuss a shared interest and plan research lessons that further it. The shared interest might be: <br> • a content area (e.g., social studies) <br> • a non-subject area (e.g., class meetings) <br> • an instructional philosophy such as "problem solving" or "whole language" <br> • an action goal, such as improved relations between ethnic groups or environmental awareness. <br> Circles meet after work hours; members decide the meeting schedule, topic, etc. |
| **Sponsor: Teachers' union** |
| **Examples:** The teachers' union may support lesson study groups in concert with or independently of local school districts; issues of social justice are one common focus. |
| **Sponsor: District** |
| **Examples:** Many districts set aside one or more afternoons a month for lesson study. Teachers join a cross-school group focused on a particular subject area (e.g., science) or educational issue (e.g., second-language learners). Also all teachers in their fifth, tenth, etc., year of teaching may gather as a cohort, in order to conduct lesson study, often focused on group-chosen topics. |
| **Sponsor: Schools (public or private) that apply for and receive funding to improve a particular area of instruction** |
| **Examples:** Schools may receive local or national funding for a specific purpose (e.g., to integrate technology into the curriculum or develop an integrated science-social studies curriculum). They accomplish this through lesson study, culminating in research lessons open to outsiders. |
| **Sponsor: University-attached laboratory schools (including "national schools" attached to public universities)** |
| **Examples:** Schools regularly conduct research lessons open to all teachers in Japan. Since the mission of these schools is to improve curriculum and instruction, thousands of teachers visit research lessons at these schools to see "what's new." |
| **Sponsor: National subject-matter associations** |
| **Examples:** Organizations such as the science, mathematics, and language arts teachers associations hold research lessons as part of their national conferences; these lessons arc designed to bring to life the members' vision of instruction, in order to stimulate further discussion and development of the vision. |

*Unless you improve your own skills, you can't do a good lesson even with a good lesson plan or good textbooks. Precisely because of this belief, we all do research lessons and try to improve our teaching skills. If you isolate yourself and do whatever you wish to do, I don't think you can ever conduct good lessons.*[30]

In addition to school-based lesson study like that at Komae School, the second common form of lesson study in Japan is district-sponsored lesson study. In many districts, all elementary teachers participate in district-sponsored lesson study, joining a group of particular interest to them that meets one or more times per month on salaried after-school time.[31] Groups may focus on instruction in a particular subject (mathematics, Japanese, art, physical education, etc.) or on a cross-cutting issue (for example, human rights education, school climate, or education of second-language learners). Typically, teachers in each group collaboratively develop a research lesson that is open to other educators in the district, and is presented on a district-wide research afternoon, when students are dismissed early except for selected classes that remain to participate in research lessons. In addition, teachers typically participate in district-based lesson study at certain intervals in their teaching careers (for example, first year, fifth year, and tenth year) when districts bring together all teachers of that cohort, using a combination of release time and after-school salaried time.

Teachers' unions, subject matter associations, and voluntary "circles" or study groups also carry on active programs of lesson study and host research lessons at the schools of their members. For example, a Japanese teachers' organization devoted to "learning through problem solving" has been influential in changing elementary instruction over recent decades through its well-attended public research lessons featuring a problem-solving approach to mathematics, science, and other areas.

Another setting for lesson study is "designated research schools" that have obtained small government grants to study new instructional approaches, such as integration of technology into the curriculum or interdisciplinary learning. Often the final product of these grants is public research lessons open to interested educators, rather than just reports submitted to the funder. Yet another common setting for research settings is Japan's 73 national elementary schools,[32] which serve a function somewhat akin to university laboratory schools in the US, and which conduct public research lessons designed to shape the future of Japanese education. Often these lessons are attended by thousands of educators who want to see cutting-edge lessons. Finally, several entities may cooperate to offer lesson study. For example, in some districts, teachers may choose their district-based lesson study from groups offered by the district, teachers' union, or subject matter associations, each of which makes curriculum materials and specialists available to the groups it supports.

Mathematics educator Takashi Nakamura points out that Japanese lesson study is not a single entity. He classifies it into three major types, according to whether its primary purpose is to:

- Solve an educational problem through development of new curriculum or instructional approaches.

- Enable practitioners to examine and improve their own practice.

- Stimulate a shared community of practice among teachers within a setting.[33]

Nakamura notes that the purpose of lesson study shapes the lesson study process — for example, the degree to which the lesson plan is jointly or individually written. Nakamura regards the second type of lesson study, focused on examination and improvement of one's own practice, as most likely to promote the development of reflective practitioners. Yet he also underlines the importance of building a shared community of practice among teachers, and he places most school-based and local lesson study in this category:

> *[The third type] of lesson study is conducted in order to create a consciousness that extends beyond individuals, and is shared throughout the school community, about issues like what kind of children we want to raise in our school community, and what vision of mathematics we want students to embrace...*
>
> *At local lesson study meetings, instructional plans are usually developed by a group, not by an individual. For example, the plan for the lower-grades research lesson is developed by the lower-grades teachers working as a group. Lesson study's function [in this case] is to create the shared consciousness that enables a group to bring to life its ideas and goals.[34]*

The vast majority of Japanese teachers participate in lesson study mainly within their own school and district, where lesson study's purpose is to develop a shared community of practice in which "everyone is involved in the effort to raise the educational level at a school."[35] But these local research lessons at ordinary local schools are enriched by the lessons conducted by national schools and subject-matter associations with the primary purpose of developing instructional approaches. For example, a lesson that includes just the right activity to help students understand a difficult mathematical concept will move rapidly from a national school or mathematics teachers' convention to schools throughout Japan, and make its way into the next round of textbooks, having been improved along the way. Although local lesson study groups in the US do not yet have an extensive network of live research lessons, videos, or reports to draw on, the public research lessons conducted by educators at Greenwich Japanese School, Paterson Public School Number Two and San Mateo-Foster City School District provide a welcome beginning.

This chapter has explored the varied settings and goals of lesson study in Japan, and examined two pioneer lesson study sites in the United States. The next chapter asks: What can teachers expect from lesson study?

# Chapter 3 Notes

[1]  Heather Crawford, Interview, March 14, 2001.

[2]  Stigler, J.W., et al. (1999). *The TIMSS videotape classroom study: Methods and findings from an exploratory research project on eighth-grade mathematics instruction in Germany, Japan, and the United States. Washington, DC:* US Government Printing Office. Stigler & Hiebert's 1999 book, *The teaching gap,* brought TIMSS video study results to a broad public audience, generating interest in prior publications such as: Lewis, C., & Tsuchida, I. (1998, Winter). A lesson is like a swiftly flowing river:  Research lessons and the improvement of Japanese education. *American Educator,* 14-17 & 50-52; Shimahara, N., & Sakai, A. (1992). Teacher internship and the culture of teaching in Japan. *British Journal of Sociology of Education,* 13, 147-162; and Shimahara, N., & Sakai, A. (1995) *Learning to teach in two cultures.*  New York: Garland.

[3]  Stigler, J.W. & Hiebert, J. (1999) *The teaching gap,* New York:  Free Press. The full case can be found in Yoshida, M. (1999). "Lesson study: A case study of a Japanese approach to improving instruction through school-based teacher development." Doctoral dissertation, University of Chicago.

[4]  A list of these efforts can be found at the website of the Lesson Study Research Group: http://www.tc.columbia.edu/lessonstudy/ or lsrg@columbia.edu.

[5]  This account of Paterson School Number Two is based heavily on Wang-Iverson, P., Liptak, L., and Jackson, W. (2000). "Journey beyond TIMSS:  Rethinking professional development."  Paper presented at the International Conference on Mathematics Education, Hangzhou, China; Research for Better Schools (2000, Fall). Against the odds, America's lesson study laboratory emerges, *RBS Currents,* 4:1, 8-10 (http://www.rbs.org); and Fernandez, C., Chokshi, S., Cannon, J., & Yoshida, M. (2001).  Learning about lesson study in the United States.  In E. Beauchamp (Ed.), *New and old voices on Japanese education.*  Armonk, N.Y.: M.E. Sharpe.

[6]  Research for Better Schools, op. cit., 8.

[7]  Comments at Lesson Study Open House, co-sponsored by Association of Mathematics Teachers of New Jersey, Paterson School Number Two, February 28, 2000.

[8]  Workshops were conducted by Frank Smith of Teachers College, Columbia University and were sponsored by the Paterson (NJ) School District.

[9]  Wang-Iverson, Liptak, & Jackson, op. cit., 12.

[10] Ibid.

[11] They used a seven-step format developed by Nanette Seago, based on the TIMSS Japanese lessons.

[12] Wang-Iverson, Liptak, & Jackson, loc. cit.

[13] Ibid.

[14] Ibid., 7.

[15] Ibid., 8.

[16] Quoted by Lynn Liptak, Presentation to School Performance Network, Pittsburgh, PA, February 1, 2000.

[17] Personal communication from James Hiebert to William Jackson, Spring 1998, subsequently emailed to Catherine Lewis, March 27, 2000.

[18] Lewis, C. & Tsuchida, I.  (1998, Winter). A lesson is like a swiftly flowing river:  Research lessons and the improvement of Japanese education. *American Educator,* 14-17 & 50-52.

[19] *The secret of trapezes* (1999). Video. 16 min. Available from lessonresearch.net.

[20] Fernandez and Yoshida's work at Paterson School Number Two was supported by funding from the National Science Foundation.

[21] Wang-Iverson, Liptak, & Jackson, loc. cit.

[22] A Japanese public school serving overseas Japanese in the New York area, which has played an important role in supporting US educators interested in lesson study.

[23] Wang-Iverson, Liptak, & Jackson, op. cit., 6.

[24] In 1998, the school's eighth graders scored above the district average on the state eighth grade mathematics exam and in the number of graduates who enrolled in Algebra 1 as high school freshman, although the school had been identified as one of the four lowest-performing of the 33 primary schools in the district when the state took over in August 1991.

[25] Jackie Hurd, Comment at lesson study planning meeting, November 1, 2000.

[26] *Video examples from the TIMSS videotape classroom study*: *Eighth grade mathematics in Germany, Japan, and the United States* (1998) is available on CD-ROM from http://nces.ed.gov/timss, publication number 98092; *Can you lift 100 kilograms?* (2000) Video. 18 min. Available from lessonresearch.net. Highlights three parts of the lesson study cycle in Japan, showing Japanese teachers engaged in planning, conducting, and discussing a fifth grade research lesson on levers.

[27] Lesson Study Research Group is at Teachers College, Columbia University, and can be reached at lsrg@columbia.edu or http://www.tc.columbia.edu/lessonstudy/.

[28] Nick Timpone, Paterson Public School Number Two, Questionnaire Response, January, 2001.

[29] Lewis & Tsuchida (1998, Winter), op. cit., 14.

[30] Shinichi Togami, Head Teacher, Inogashira Elementary School, Musashino, Japan, Interview, July 2, 1996.

[31] Working hours are a negotiated item and vary by district, but the official work day for Japanese elementary teachers typically ends between 4:30 and 5:30 pm.

[32] These national public elementary schools (*kokuritsu fuzoku shougakkou*) are university-affiliated laboratory schools whose mission is to improve education.

[33] Nakamura, T., quoted in *Zadankai: Shougakkou ni okeru jugyou kenkyuu no arikata wo kangaeru.* (Panel Discussion: Considering the nature of lesson study in elementary schools) 14-15, in Ishikawa, K., Hayakawa, K., Fujinaka, T., Nakamura, T., Moriya, I., & Takii, A. (2001). *Nihon Suugaku Kyouiku Gakkai Zasshi (Journal of Japan Society of Mathematical Education)*, 84:4, 14-23 (in Japanese).

[34] Ibid., 15.

[35] Takii, quoted in Ishikawa, et al., Ibid*.,* 16.

# 4

# What Can Teachers Expect from Lesson Study?

> Lesson study keeps me on my toes, keeps me from getting complacent, keeps me challenged professionally. I compare it to the feeling I get when I am taking a graduate course. Lesson study allows me to build relationships with my colleagues. Lesson study opens the classroom doors and minimizes isolation.
>
> - Nick Timpone, Teacher, Paterson Public School Number Two[1]
>
> What's the most important benefit of lesson study? You develop the eyes to see children.
>
> - Kyouichi Itoh, Elementary Principal, Nagoya, Japan[2]

Before delving into the "how-to" of lesson study (the focus of the next chapter), it makes sense to ask whether lesson study is right for you and for your school. This chapter describes the experiences you can expect as part of lesson study. The items listed in Figure 5, which is designed to guide reflection after lesson study, summarize the opportunities that well-designed lesson study should provide. But now, before you commit time to lesson study, is an even better time to consider the opportunities lesson study should provide. This chapter briefly discusses each of the eight experiences outlined in Figure 5.

**Figure 5
Reflection on Lesson Study**

---

Has lesson study enabled us to:

- Think carefully about the goals of a particular lesson, unit, and subject area?

- Study and improve the best available lessons?

- Deepen our subject-matter knowledge?

- Think deeply about our long-term goals for students?

- Collaboratively plan lessons?

- Carefully study student learning and behavior?

- Develop powerful instructional knowledge?

- See one's own teaching through the eyes of students and colleagues?

---

## Think Carefully about the Goals of a Particular Lesson, Unit, and Subject Area

Lesson study gives teachers a chance to think in depth about a particular lesson, unit, and subject area.[3]  As one Japanese teacher said:

> *Research lessons are very meaningful for teachers because ...we think hard and in a fundamental way about several critical issues.  For example: What is the basic goal of this lesson in this textbook? How does this particular lesson relate to my students' learning and progress in this school year?  How does this lesson relate to other curriculum areas?   Thus, it is very beneficial to teachers.  Unless we think about all these things, we can't conduct research lessons.  That is the purpose or significance of research lessons.  Even if teachers do not think hard about the lessons they teach daily from the textbook, they must really rethink the fundamental issues for research lessons.[4]*

When choosing the academic subject and topic for lesson study, teachers often:

- Target a weakness in student learning or development.

- Choose a topic teachers find difficult to teach.

- Choose a subject that has changed recently, for example, new content, technology, or teaching approaches that have been advocated.

- Concentrate on the study of Japanese and mathematics in alternate years, since these subjects account for much instructional time and can be fundamental to progress in other areas.

As Chapter 6 explores in detail, lesson study is *not just about a single lesson*, but about the teaching of an entire unit and subject area, and indeed, about student development more broadly.  The academic content goals of lesson study are often taken directly from the *Course of Study*.[5]  For example, the Komae teachers drew from the *Course of Study* the three things about levers that students were expected to learn during the levers unit (see instructional plan, Appendix 2).  In contrast, the Komae teachers wrote the research theme themselves, based on careful faculty-wide discussion of their ideals for students.

A US teacher described the impact of lesson study on his instruction:

> *The most notable change in my lesson planning and teaching has been the questions that I ask myself. The first question I ask myself about a lesson is "what do I want the students to learn from this lesson?"  While this may seem an obvious question to ask, it was never something I asked myself until I began the lesson study process. The question I*

*was asking myself before lesson study was more like "what am I covering today?"*[6]

The instructional plans in Appendices 2 through 4 provide examples of the kinds of lesson, unit, and subject area goals that are discussed during lesson study. Developed by educators in different regions of Japan and by different types of lesson study groups, they illustrate a range of formats for instructional plans.

## Study and Improve the Best Available Lessons

When Japanese teachers plan research lessons, they draw on the best ideas from inside and outside the school, and devote much more time to lesson planning and discussion than is usually possible in the hubbub of daily school life. In *Can You Lift 100 Kilograms?*,[7] teachers began their lesson planning by comparing several different unit plans for teaching levers, which were drawn from textbooks, teachers' own prior instruction, and research lessons that members of the group had observed or found in books written by teachers. Like the teachers at Komae, other Japanese teachers can easily locate instructional plans for units ranging from poetry appreciation to division of fractions. Any large Japanese bookstore has hundreds of lesson study volumes written by teachers, explaining their long-term goals and instructional philosophy, and providing lesson and unit plans, examples of student work, reflections on the strengths and difficulties of the lesson, and practical guidance for teachers who want to try the lessons. As other teachers draw on these lessons, and add, test, and report their own improvements, the quality of lessons steadily improves.

Obviously, US teachers do not yet have access to the wide range of lessons available to their Japanese colleagues. But by improving on and sharing the best currently available lessons, US teachers can begin this process. Teachers engaged in lesson study can see lessons that represent the best wisdom of their colleagues, and can hear the reactions of other teachers. Lessons naturally improve as the ideas of many teachers are brought to bear. To quote a Japanese proverb: "When you gather three people, you have a genius."

## Deepen Our Subject-Matter Knowledge

A friend of mine humorously recalls what it felt like to learn science during the first year of medical school:

> *Every week a renowned basic researcher would lecture us. It felt like each researcher filled up a huge dump truck with all the information that a future doctor might possibly need, backed it up to the lecture hall, and dumped it on the medical students. Minutes into each lecture, I fell asleep. Later I learned what I needed to know, not by crawling around to*

*sift through all the dumped information, but by strategically figuring out what I needed to know to solve particular patients' cases and to understand particular diseases.*[8]

His account points up a problem that doctors and teachers share as clinicians: the danger of being overwhelmed by research and theory of unknown clinical relevance. Lesson study can help teachers in the way the patient cases helped the medical student, by helping to identify and organize the information needed to solve a problem.

When solar energy was added to the Japanese science curriculum about a decade ago, teachers in many schools chose to focus research lessons on it. They debated among themselves what knowledge and attitudes related to solar energy were important for students to develop. For example, during the discussion following a research lesson, a teacher said:

> *I haven't taught fourth graders for awhile; so I have no idea how and why solar batteries were added to the curriculum. I'm only guessing that including solar batteries reflects adults' hope that children will become the next generation of scientists who will become interested in solar energy and thereby help Japan. Science education specialists might be concerned about children using the proper vocabulary or setting up certain experimental conditions, but if the goal of including solar batteries in the curriculum is to get children interested in the fact that electric current can be changed by light, then Mr. H.'s lesson fulfilled that. So I'd really like to know the reason why solar batteries were included as a new curriculum material for fourth graders.*[9]

Teachers had the benefit of colleagues' ideas as they considered what it was important for students to understand about the new science content. The teachers also invited outside science educators to the research lessons, and questioned them about the new science content. One teacher asked:

> *I want to know whether the three conditions the children described, "to put the solar battery closer to the light source," "to make the light stronger," and "to gather the light" would all be considered the same thing by scientists. They don't seem the same to me. But I want to ask the teachers who know science whether scientists would regard them as the same thing.*[10]

In other words, the research lesson provided an opportunity for these teachers to collectively figure out what knowledge was important, discover gaps in their own understanding, and acquire the needed information.

As they work together on study lessons, US teachers also engage in conversations likely to deepen content knowledge. For example, as one group of US teachers imagined the various types of triangles students might construct during a lesson, they raised and discussed several questions: "Is a scalene triangle always obtuse?" "Does a right triangle have two equal sides?" One teacher commented that these questions arose because their images of particular triangles tended to get "locked into" the textbook's pictures.[11]

The potential of lesson planning to stimulate development of teachers' content knowledge is also suggested by research from China, where teachers develop, discuss, and refine a shared set of examples for conveying key mathematical ideas in a process of public teaching that shares some features with lesson study.[12] Chinese elementary teachers take fewer courses in advanced mathematics than do US teachers, but some evidence suggests they are better able to provide concrete examples of fundamental mathematical ideas, for example, to devise a word problem that illustrates what it means to divide a fraction by another fraction.[13] Such exploration of fundamental mathematics may be more likely to occur during processes like lesson study (as teachers develop problems and explore student thinking) than during the higher-level mathematics courses that are often proposed as a way to boost subject-matter knowledge.

As they anticipate and analyze student thinking, teachers engaged in lesson study discover gaps in their own understanding, and seek out information in many ways: from colleagues, from print and video resources, from local subject specialists, and from knowledgeable outsiders, such as university-based researchers and technical assistance providers (see Figure 6). A lower-grade teacher at Paterson School Number Two comments on lesson study's role in building her subject-matter knowledge:

> *We've gotten into a lot of discussions from [reading] Liping Ma and [thinking about] the knowledge piece — what students really need to take with them and why they are doing this... And we are becoming more educated... The seventh grade teachers... throw stuff back at us. We are relearning the mathematics that we learned when we were in school that some of us have forgotten. So I think our knowledge has increased as we have been doing this.[14]*

Lesson study alone does not ensure access to content knowledge. But teachers are likely to build their content knowledge as they study good lessons, anticipate student thinking, discuss student work with colleagues, and call on outside specialists. Lesson study can help educators notice gaps in their own understanding and provide a meaningful, motivating context to remedy them. As Liping Ma observed, "American educators assume that you need to know content knowledge before you can plan lessons. Chinese teachers think you learn content knowledge by planning lessons."[15]

## Think Deeply about Our Long-Term Goals for Students

Lesson study starts when teachers agree upon a shared goal for improvement, usually called the "research theme" or "important aim."[16] In the case of school-based lesson study like that of the Komae teachers featured in Chapter 1, this shared goal is developed when teachers

**Figure 6**
**The Role of Outside Specialists in Japanese Lesson Study***

The central force in a research lesson is the teacher or group of teachers who develop the lesson. Observing teachers also play an important role. A research lesson would not function as professional development without these two groups of players. A third player who sometimes participates is the invited outside specialist. For mathematics lesson study, this role may be filled by a teacher, principal, or university mathematics educator who is known for his or her expertise in mathematics teaching. The invited specialist may comment on the lesson, provide advice as the lesson is developed, teach a research lesson, or provide a summary at the faculty colloquium.

Professor K is an experienced mathematics teacher educator at a national university. He is often invited to comment on research lessons. As a rule of thumb, he tries to "praise ten and criticize one." In other words, he selects one (or a few) important ideas to focus on and chooses not to dwell on other areas needing improvement.

When he is invited to comment on a research lesson, Professor K carefully considers the context. For example, at a lesson he tries to focus on generalizing the main idea of the research lesson. Since no two classrooms are the same, the outside commentator must point up the good practices exemplified in the research lesson and help teachers think about adaptation of the practices to their classrooms.

Occasionally, an outside commentator works closely with a school or a group of teachers, attending a number of research lessons conducted by the group over an extended period of time. For example, Professor S has been working with a public elementary school in the Tokyo area for over two years, attending their in-house lesson study days as the outside commentator. Three years ago, the school was struggling with many problems when a new principal arrived. She wanted to rejuvenate their lesson study and she invited Professor S to be their outside commentator. Since then, Professor S has worked with the faculty members to develop a lesson study program at the school. After a recent lesson study open house, which was attended by more than 200 teachers, Professor S recalled the initial lesson study meetings at the school. During those meetings, more than half of the faculty were either asleep or pretended to be. However, the principal, Professor S, and the head of the school lesson study group persisted. Today, all teachers at the school seem to have found joy and excitement in lesson study. During the two years he has been involved, Professor S shared his expertise and was also a cheerleader who encouraged the teachers to keep moving on. He even taught a research lesson himself.

Outside specialists can play a number of different roles. Something common to all effective specialists is that they pay attention to the audience and anticipate what they are ready to learn. In that sense, an outside specialist is just like a classroom teacher. Just as a teacher must assess and respond to students' needs, the outside commentator must do the same with the teachers attending the lesson study. As more US educators try to implement lesson study, we must pay attention to leadership training for outside specialists who work with lesson study groups.

*by Tad Watanabe

consider the ideal qualities they hope students will have when they graduate, their current qualities, and the gap between the two.[17] Japanese teachers usually choose a broad goal that is compelling to teachers from all grade levels and many points of view, as the following lesson study goals from Japanese elementary schools illustrate:

- to develop instruction that ensures students' basic academic abilities, fosters their individuality... [and] meets students' individual needs[18]

- for students to take pleasure in friendships and learning

- for our instruction to be such that students learn eagerly.[19]

By focusing on the qualities they hope students will have several years down the road, Japanese teachers try to choose a goal that is worthy of sustained, long-term work, rather than faddish, trivial, or tied to narrow outcomes such as test scores.

Americans often find puzzling the relationship between lesson study and long-term goals for students. A few eyebrows always go up when I read the lesson study theme from Komae School Number Seven: "For students to value friendship, develop their own perspectives and ways of thinking, and enjoy science."[20] The skeptical faces seem to say, "What on earth do friendships have to do with learning science anyway? Let's skip the fuzzy stuff." Since the importance of concrete and measurable outcomes is so often hammered into US educators, the Japanese focus on broad and long-term goals, as well as immediate learning, can be puzzling. But to many US educators, the opportunity to consider long-term goals feels like the essential missing piece of instructional improvement. As one US teacher commented, "Lesson study focuses on the long-term; usually when you're teaching you don't have time to think beyond the immediate skills you want students to learn that day."[21] Another US teacher said:

> *A lot of [American] schools develop mission statements, but we don't do anything with them. The mission statements get put in a drawer and then teachers become cynical because the mission statements don't go anywhere. Lesson study gives guts to a mission statement, makes it real, and brings it to life.[22]*

Building lesson study around long-term goals may offer protection against faddism and trivial goals. The long-term focus enables Japanese teachers to keep firmly in mind the qualities such as love of learning and capacity to get along with others that can otherwise get lost in the daily grind of school, a benefit that may be especially important in the US, where high-stakes tests can easily eclipse long-term thinking. Lesson study's long-term goals recognize that learning is greatly shaped by students' curiosity, motivation, sense of support from classmates, and other qualities of heart and mind. In their levers lesson, Komae teachers gathered data on outcomes including students' "shining eyes" and *tsubuyaki* (underbreath exclamations or "aha's"). When I ask a roomful of US educators whether they think shining eyes and aha's are important to students' future science learning, the answer is generally a resounding "yes."

The question that animates whole-school lesson study — "What qualities do we hope our students will develop?" — is a question that is likely to be deeply motivating to every teacher. It provides a way to connect daily lessons to our most cherished long-term goals for students. While we all appreciate the importance of teaching students to write a good paragraph and multiply fractions, what drew most of us into teaching was something broader: the opportunity to shape the next generation and, indeed, the future society in which we will live. To the extent that lesson study's long-term goals connect us to that larger educational purpose, they will provide motivational fuel. A US student teacher noted, " I really like how lesson study is connected to a larger goal; even the tiny details of the lesson don't seem mundane, because they are connected to a larger goal."[23]

## Collaboratively Plan Lessons

"Isolation is the enemy of improvement," notes educator Richard Elmore.[24] While the average Japanese teacher sees about ten research lessons a year,[25] US teachers have few opportunities to observe lessons taught by others.[26] A 33-year veteran of US elementary teaching routinely advises her student teachers: "Observe as many teachers as you possibly can during your student teaching year. This is the last time in your career you will have the chance!"[27]

Even if US teachers invest time in collaboration, not every collaborative experience yields dividends in the classroom. A US teacher commented, "I've been part of a study group for years, but I don't know what the impact has been on my class."[28] Because lesson study focuses on shared practical work to improve lessons, it has the potential to build collaboration while at the same time yielding dividends for students.

Lesson study also provides a natural way for teachers to think about how their own teaching connects with what other teachers are doing. Teacher Heather Crawford reflects on lesson study's impact at Paterson School Two:

> In the past, a lot of us never really thought about two grades down the line and how what we were teaching affects them. And now we really are. We are looking at it from [the point of view of] "This is what they learn in kindergarten. How does it carry through eighth grade?[29]

Japanese teachers view collaboration as an important benefit of lesson study. As one said, "What's a successful research lesson? It's not so much what happens in the research lesson itself that makes it successful or unsuccessful. It is what you learned working with your colleagues on the way there."[30]

## Carefully Study Student Learning and Behavior

During research lessons, teachers scour the classroom for evidence of student learning, motivation, and behavior. Figure 7 provides examples of the kinds of data Japanese teachers

**Figure 7**
**Data Collection during Research Lessons: Examples of Focal Questions**

| **Type of Outcome: Academic Learning** |
| --- |
| Did students notice a pattern in the number combinations to break apart 10? |
| How many students shifted from simple counting to a more flexible method of subtraction with regrouping? |
| How many students spontaneously designed controlled experiments? |
| Could students design three experiments on levers, justify their experimental designs, and predict the results? |

| **Type of Outcome: Motivation and Engagement** |
| --- |
| Did students' tone of voice, body language, persistence, exclamations under breath, "aha" comments, "shining eyes," etc., provide evidence of motivation and engagement? |
| Did students try multiple ways to solve the problem? |
| What were the quality and quantity of student writing in journals? |
| What proportion of students raised their hands, volunteered their ideas in whole-class discussion, and spoke up in small groups? How many boys and girls participated in each of the above? |

| **Type of Outcome: Social Behavior** |
| --- |
| How evenly did students share speaking time within their small groups? |
| How many times did students refer to one another's comments and build on them? |
| When students disagreed with one another, was the tone respectful or dismissive? |
| How often did the five students who are normally very quiet speak up during this lesson? How did their classmates respond? |

| **Type of Outcome: Student Attitudes Toward Learning** |
| --- |
| What was the participation by three selected students who have very different levels of science achievement and interest? |
| Did students find this lesson more or less interesting than their usual science lessons? What was best and worst about it? (assessed by questionnaire) |
| What would students change about this lesson next year? What did they learn, and how do they like the topic? |

| **Type of Outcome: Instructional Features, Information Requested by Instructor** |
| --- |
| What time did each component of the lesson (e.g., introduction, problem posing, individual work, group work, class discussion) begin and end? |
| What questions did the teacher ask? How did the teacher respond to student answers? Which students were called on by the teacher? |

collect. Each group member has a data collection assignment — for example, to record changes in student thinking or all the questions students ask. Data collected typically include evidence of learning, of interest or motivation, and of students' treatment of one another, reflecting the belief that interest and classroom climate, as well as academic knowledge, are important predictors of future learning. Teachers may collect data from the overall unit as well as the research lesson itself.

Data collection enables teachers to see instruction through the eyes of the students. What is their understanding of the subject matter? Are students pumped up with the thrill of discovery or just grudgingly going through the motions? Are they engaging or ignoring each other's ideas? A US teacher commented that lesson study "takes reflection to the next level; it makes it practical and tangible."[31] The gold standard for judging the research lesson is student learning and development. Techniques such as cooperative grouping, technology, or manipulatives do not become ends in themselves. James Stigler and James Hiebert write:

> *Reform documents that focus teachers' attention on features of "good teaching" in the absence of supporting contexts might actually divert attention away from the more important goals of student learning. They may inadvertently cause teachers to substitute the means for the ends — to define success in terms of specific features or activities instead of long-term improvements in learning.*[32]

Theory and research provide excellent starting points for understanding what good instruction is, and, in a perfectly controlled world, the "best practices" documented by research might be the same in every classroom. But in the real world, every class is different. Lesson study assumes that teachers need to look for evidence of students' learning, motivation, and development in their own classrooms. Lesson study also provides a means for teachers to develop their evidence-gathering skills and ability to see a lesson from the student's point of view. Developing "the eyes to see children" is, in the view of many Japanese educators, the most important goal of lesson study.

## Develop Powerful Instructional Knowledge

Lesson study also builds understanding of instructional techniques, as teachers hone lessons to better reach children. Teachers studied by Makoto Yoshida developed a list of characteristics of a good manipulative, after discovering that the manipulatives used to teach subtraction did not allow teachers to reconstruct students' thinking.[33] The teachers of *Can You Lift 100 Kilograms?*[34] discovered that students gave very different responses when asked to look at an actual 220-pound sack than when given an illustration of the sack on a worksheet. In lesson study, teachers think carefully about the questions, activities, manipulatives, and visual aids to be used in the lesson. By observing students, teachers see how a particular problem or approach animates or derails learning.

Through lesson study, teachers develop and improve teaching strategies that can be applied throughout the curriculum, such as how to pose a good *hatsumon* (major question or problem) that will sustain students' interest throughout the lesson and unit, how to use debates to maximize student participation in discussions, and how to foster student note-taking and reflection. In Appendix 1, Makoto Yoshida writes about the art of blackboard writing that has spread through lesson study. As teachers watch and analyze research lessons, they develop a shared vocabulary for talking about the fine points of teaching and learning. For example, a Japanese dictionary of lesson study defines several hundred terms related to classroom practice.[35] Japanese teachers can name many techniques they have picked up through participation in research lessons, such as putting student names on magnets (so students can register on the board their agreement or disagreement with ideas under discussion), and having students use hand signals to show whether they want to agree, disagree, or make a new point.

Beyond specific techniques, teachers may find that research lessons affect their whole philosophy of teaching:

> *I had always seen education as teachers giving knowledge to children, as a top-down process. Through my work with the elementary science research group, I came to see education not as giving knowledge to children but as giving them opportunities to build their own knowledge. Initially, that was not what I believed. Even when I saw it in practice, I couldn't believe in it at first. When I first saw lessons in which children were building their own knowledge, I thought "Is this kind of instruction really OK? It takes so much time." But then I began to realize that if children don't have experiences in which they personally give birth to knowledge, they don't understand the meaning of that knowledge. Students who lack these experiences can memorize the knowledge, but when the time comes to use it, they can't.*[36]

## See One's Own Teaching through the Eyes of Colleagues and Students

Not every member of a lesson study group will teach a research lesson every year; even in Japan opportunities to teach research lessons occur infrequently (once a year or so).[37] But for teachers who do teach a research lesson, the data gathered by colleagues provide a mirror on one's own teaching. Colleagues can, for example, record the discussion in every small group, count how many boys and girls raise their hands, or record data of particular interest to the instructor, such as all the questions or responses by the instructor. Often a Japanese teacher will select three students who differ in achievement (or other target characteristics) and ask colleagues to collect data on the lesson and unit from the point of view of those students, keeping records of all their interactions, written work, and so forth. In this way, teachers can see how the lesson was experienced by students with varied strengths and challenges. As a Japanese teacher eloquently notes, research lessons provide a mirror on one's practice:

*A lesson is like a swiftly flowing river; when you're teaching you must make judgments instantly. When you do a research lesson, your colleagues write down your words and the students' words. Your real profile as a teacher is revealed to you for the first time.[38]*

# Summary

Figure 8 lists qualities of effective professional development identified by research. Lesson study exemplifies these qualities. It occurs in a real, motivating context (the classroom) and focuses on a problem of great interest to teachers: their hopes for student learning and development. The research lesson provides an ongoing method to improve instruction, ensuring that good ideas are not just talked about, but brought to life. To plan research lessons, teachers draw on expertise from within and outside the school; they gather the best lessons and instructional techniques from the nation (or world) and improve these through careful observation of their own students. Lesson study builds collaboration as teachers progressively improve lessons that are "our" lessons, not "my" lessons.

**Figure 8**
**Qualities of Effective Professional Development**

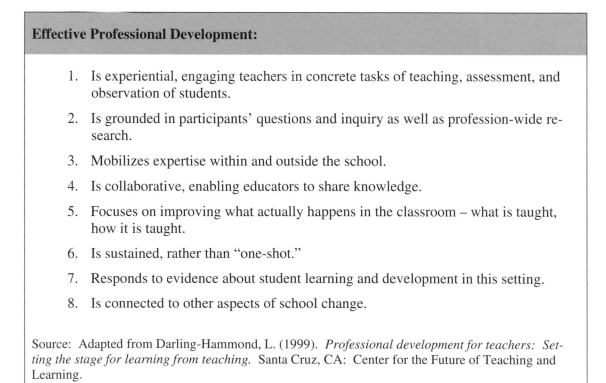

**Effective Professional Development:**

1. Is experiential, engaging teachers in concrete tasks of teaching, assessment, and observation of students.

2. Is grounded in participants' questions and inquiry as well as profession-wide research.

3. Mobilizes expertise within and outside the school.

4. Is collaborative, enabling educators to share knowledge.

5. Focuses on improving what actually happens in the classroom – what is taught, how it is taught.

6. Is sustained, rather than "one-shot."

7. Responds to evidence about student learning and development in this setting.

8. Is connected to other aspects of school change.

Source: Adapted from Darling-Hammond, L. (1999). *Professional development for teachers: Setting the stage for learning from teaching.* Santa Cruz, CA: Center for the Future of Teaching and Learning.

# Chapter 4 Notes

[1] Nick Timpone, Questionnaire Response, January, 2001.

[2] Kyouichi Itoh, Interview, August 15, 2001.

[3] In Japan, lesson study may be conducted in any "academic" or "nonacademic" subject area (e.g., mathematics, Japanese, social studies, science, class meetings, physical education, art, music, school climate).

[4] Japanese elementary teacher, Gakugei Fuzoku Setagaya Elementary School, Interview with Ineko Tsuchida, February 2, 1996.

[5] *Course of study for elementary schools* is an English title for the Japanese Ministry of Education's *Shougakkou gakushuu shidoo youryou* (Tokyo: Government of Japan, 2000), a half-letter size booklet of 100 pages that lays out the minimum hours, goals and content for all 12 required content areas (Japanese, social studies, mathematics, science, life environment studies, music, art, homemaking, physical education, ethics, special activities, and interdisciplinary studies).

[6] Timpone, loc. cit.

[7] *Can you lift 100 kilograms?* is an 18-minute video of the lesson study cycle in a Japanese school, available from lessonresearch.net.

[8] Andrew Leavitt, Personal Communication, December 20, 1978.

[9] Lewis, C. & Tsuchida, I. (1998, Winter). A lesson is like a swiftly flowing river: Research lessons and the improvement of Japanese education. *American Educator*, 16.

[10] Teacher's Comment, Research Lesson Discussion, Sannou School, Ohta-ku, Tokyo, June 25, 1996.

[11] Perry, R., Lewis, C., & Akiba, M. (April 2, 2002). Lesson Study in the San Mateo-Foster City School District. Paper presented at Annual Meetings of the American Educational Research Association, New Orleans, LA.

[12] Wang, J. (March 15, 2001). Learning to teach with mandated curriculum and public examination of teaching as contexts. Paper presented at the annual meeting of Comparative and International Education Society, Washington, DC.

[13] Ma, L. (1999). *Knowing and teaching elementary mathematics: Teachers' understanding of fundamental mathematics in China and the United States.* Mahwah, NJ: Lawrence Erlbaum Associates.

[14] Heather Crawford, Interview, March 14, 2001.

[15] Ma, L. (April 23, 2001), Remarks at Conference "Learn globally, teach locally," Mid-Atlantic Eisenhower Consortium for Mathematics and Science Education at Research for Better Schools, Cherry Hill, NJ.

[16] *kenkyuushudai, kenkyuuteema, juutenmokuhyo.*

[17] For a description of this process in a US school, see: Fernandez, C., Chokshi, S., Cannon, J., & Yoshida, M. (2001). Learning about lesson study in the United States. In E. Beauchamp (Ed.), *New and old voices on Japanese education.* Armonk, N.Y.: M.E. Sharpe.

[18] This goal is from Yoshida, M. (1999). "Lesson study: A case study of a Japanese approach to improving instruction through school-based teacher development." Doctoral dissertation, University of Chicago, 51. The example is edited to conform with Japanese translations used in this volume.

Yoshida provides additional examples of lesson study goals in the dissertation, in press as: Yoshida, M. & Fernandez, C. *Lesson study: A case study of a Japanese approach to improving instruction through school-based teacher development.* Mahwah, NJ: Lawrence Erlbaum.

[19] From elementary research lessons in Shizuoka and Tokyo, Japan, July, 1998.

[20] *Can you lift 100 kilograms?*, loc. cit.

[21] Class discussion comment by first-year teacher, Mills College class, Oakland CA, January 16, 2001.

[22] Class discussion comment by first-year teacher, Mills College class, Oakland CA, January 16, 2001.

[23] Class discussion comment by student teacher, Mills College class, Oakland, CA, January 23, 2001.

[24] Elmore, R. (1999-2000 Winter). Building a new structure for school leadership. *American Educator*,12.

[25] Yoshida (1999), op. cit.

[26] Ibid.; Darling-Hammond, L. & Ball, D.L. (1998). *Teaching for high standards: What policymakers need to know and be able to do.* New York: National Commission on Teaching and America's Future and Consortium for Policy Research in Education; and Darling-Hammond, L. (1997). *The right to learn: A blueprint for creating schools that work.* San Francisco: Jossey-Bass.

[27] Katie Johnson, Teacher, Oxford Elementary School, Berkeley, CA, Comment during lesson study planning meeting, February 20, 2001.

[28] I am indebted to Patsy Wang-Iverson for passing on this quote.

[29] Crawford, loc. cit.

[30] Lewis, C. (2002). Does lesson study have a future in the United States? *Nagoya Journal of Education and Human Development,* 1, 11.

[31] Class discussion comment by first-year teacher, Mills College, January 16, 2001.

[32] Stigler, J.W. & Hiebert, J. (1999). *The teaching gap.* New York: Free Press. The full case can be found in Yoshida, M. (1999). "Lesson study: A case study of a Japanese approach to improving instruction through school-based teacher development." Doctoral dissertation, University of Chicago.

[33] Yoshida (1999), op. cit.

[34] *Can you lift 100 kilograms?,* loc. cit.

[35] Yokosuka, K. (1996). *Jugyou kenkyuu yougou jiten* (*Dictionary of lesson study terms*). Tokyo: Kyoiku Shuppan Kabushiki Kaisha.

[36] Kazuyoshi Morita, teacher at Tsukuba attached elementary school, Interview, July 3, 1996.

[37] Yoshida (1999), op. cit.

[38] Morita, loc. cit.

# 5

# Time and Scheduling

**I would advise teachers who are just beginning to consider lesson study to forge ahead slowly. Lesson study is not something that one can "jump into." Understand what it entails. Don't skip any steps and be prepared to give and take "constructive criticism" in the true sense of the phrase. Understand that lesson study is about the process as much as it is about the lesson.**

**- Nick Timpone, Teacher, Paterson Public School Number Two[1]**

How much time does lesson study require, and where can US teachers find it? What is a good schedule for lesson study? Time and scheduling for lesson study are often the first questions raised by US educators. For that reason, this chapter briefly outlines the time needed for each phase of lesson study and explores issues related to scheduling, including whether videotaping can be substituted for live observation of research lessons. This chapter is followed by a step-by-step discussion of lesson study (Chapter 6).

## Lesson Study: The Basic Sequence of Activities

Lesson study consists of three basic activities: collaborative planning, the research lesson, and discussion/revision of the research lesson. Figure 9 shows the sequence of these activities and estimates the number of meetings (of 45-90 minutes) needed for each. Just how much time each step will take depends upon your particular group — your goals, familiarity with each other and with lesson study, whether you have good curriculum materials appropriate to your goals, and whether your research lesson will be taught and revised once, twice or three times. The thick-bordered steps at the bottom of the figure are optional, and may be repeated as many times as desired. The benefits of re-teaching lessons are laid out in Figure 10.

Optimal group size for planning a research lesson is probably about four to six teachers. But for activities like selecting your research theme and observing and discussing lessons (particularly on the second or third teaching), you may benefit by working together with other groups. For school-based lesson study, teachers from one or two adjacent grade levels typically form a group.

## Figure 9
## Lesson Study Schedule

| Number of Meetings | Task | Participants | Activities |
|---|---|---|---|
| 1-3 | Select research theme, subject area | All groups together | Agree on research theme, based on discussion of long-term goals for students. Select subject area (e.g., mathematics). |
| 3-6 | Plan research lesson | Research lesson planning group (4-6 members) | Select topic for research lesson. Outline the unit and plan the research lesson. Write thorough instructional plan, as described in Chapter 6. |
| 1 (a class period) | Conduct research lesson | Research lesson planning group, other invitees as desired | One member teaches research lesson; others observe and collect data agreed upon in advance. |
| 1 | Discuss research lesson | Same as prior step | Discuss data from research lesson soon after the research lesson (same day). |
| 1-2 | Reflect-revise | Research lesson planning group | Consolidate what was learned from research lesson and write up reflections. If desired, revise lesson for re-teaching. |
| 1 (a class period) | Second teaching of research lesson | Research lesson planning group and invitees or whole faculty as desired | A second group member re-teaches research lesson to own class; others observe and collect data agreed upon in advance. |
| 1 | Discuss research lesson | Research lesson planning group and invitees or whole faculty as desired | Discuss second teaching of research lesson soon after the research lesson (same day); revise lesson again if desired. |
| 1-2 | Reflect-revise | Research lesson planning group | Reflect on lesson study effort and goals, and continue or modify these. |

Note: Bolded box contains optional steps that may be repeated one or more times.

**Figure 10**
**Why Re-teach Lessons?\***

Especially when preparing a lesson study open house for the public, Japanese teachers often re-teach a lesson after revising it. Why? I'd like to share some ideas about re-teaching from Japanese educators.

Re-teaching lessons gives teachers within a school more opportunities to teach in front of others, and to observe various lessons, classrooms, and students. Re-teaching a lesson also enables teachers to observe the impact of their discussion and subsequent revision of the lesson. After teaching a research lesson once, teachers can discuss the lesson in a more concrete fashion and on a deeper level, and can approach its revision in a more thoughtful and systematic manner, based on actual observation of the students' learning process. Lesson observation and discussion with experienced teachers are particularly important for novice teachers, who need to develop good "eyes" to observe lessons and skills to reshape lessons to maximize student understanding.

Teachers from the Greenwich Japanese School in Connecticut and Paterson Public School Number Two in New Jersey voluntarily established a Math Lesson Study Circle in July 2001 and have been meeting at least once a month for joint lesson study. During these meetings, when the group is having difficulty resolving some issue in lesson design, the Japanese teachers have sometimes suggested teaching the lesson as it currently is, and using the information from teaching the lesson to revise it. They have also mentioned that re-teaching in a second classroom by a second teacher creates a totally different environment. Since there are no perfect lessons that are successful for every classroom situation, learning how to adjust a lesson is a valuable skill for teachers.

Why re-teach a lesson? According to Japanese educators, re-teaching increases opportunities for teachers to learn from one another, reveals the power for students of the revised lesson, and teaches valuable skills of lesson observation, discussion, and adaptation that are fundamental to improvement of teaching.

\*by Makoto Yoshida

## Can Videotape Be Substituted for Live Observation of Research Lessons?

Videotape offers certain advantages over live observation of research lessons, such as flexibility of scheduling and the possibility of repeated viewing. Some teachers feel more comfortable being videotaped (with the knowledge that they can later choose not to share the video) than being observed live. But live research lessons are the heart of lesson study in Japan, and teachers sometimes travel hundreds of miles to attend them. Why do Japanese teachers accord so much importance to live observation?

When teachers watch a research lesson, they notice many things that cannot be gathered from student tests and written work, or sometimes even from videotapes. For example, teachers study students' engagement, persistence, interactions within small groups, and *tsubuyaki* (under-breath exclamations or aha's). During research lessons teachers observe students' whole demeanor toward learning and toward one another.

Although educators new to lesson study sometimes imagine that lesson plans will capture the essence of their lesson study work, student learning and development cannot be assessed by looking at a lesson plan. The "Tale of Two Lessons" (Figure 11) illustrates the danger of trying to identify a good lesson without actual observation. To say "It was a good lesson but the students didn't get it," is like saying "the operation was successful but the patient died."

Videotape, audiotape, lesson plans, photographs, and student work are all used extensively by Japanese teachers to document research lessons. But they are not regarded as a substitute for live observation of research lessons, when teachers can actively record the participation of all students, scan the room for evidence that students understand a task, closely study the work of a small group of students, and pick up the general mood and interest level in the classroom. Videographers must decide (generally in advance of the lesson) where to focus the camera(s), and this inevitably narrows the stream of experience captured on tape. In contrast, live observation enables teachers to follow the "swiftly flowing river" of instruction in unanticipated directions, and to pick up the mutterings and shining eyes of the students.

## Should Groups Meet during or after School?

Most lesson study meetings can take place either during or after school hours, depending on your preference. But since the research lesson itself includes students, it generally takes place during school hours and requires the observing teachers to be released from their own teaching responsibilities for the period of the lesson. (Some schools arrange in advance for one class to stay after school for the research lesson, or for all but one class to be dismissed early.) Another way to free up one class for a research lesson is to integrate your lesson study plan with a scheduled ongoing school-wide activity (such as an art project, community service project, or science fair) that students participate in regularly over the course of a year, with just the students from the research lesson class missing this activity on a rotating basis. Often such activities can enlist parent and community volunteers, specialist teachers from outside the school, artists in residence, or other adults who can free up classroom teachers while providing a memorable learning experience for children. Books like *At Home in Our Schools*[2] provide suggestions for such school-wide events.

Meetings that include just teachers from one or two grade levels may be scheduled during a common planning time, or at a time when the principal, assistant principal, or another staff member can teach these classes. This can be a wonderful opportunity for school or district administrators to demonstrate their commitment to teachers' professional development and retaste the flavor of classroom life!

**Figure 11**
**A Tale of Two Lessons**

Several weeks apart, I saw the "same" probability lesson in two US fourth-grade class-rooms. The lesson's basic plan is:

- Working in pairs, students draw ten marbles from a large sack of marbles. Each pair of students then predicts the proportion of black and white marbles in the big sack based on their own sample of ten marbles.

- Students look at the data from all the pairs of students, and decide whether to revise their own predictions.

- Students discuss whether using the data from all the pairs is likely to lead to a better prediction than using only their own data, and why or why not.

- Students count the marbles in the sack, and see whether their individual prediction or the group average came closer to the actual number.

In the fourth grade classroom where I first saw this lesson, it worked like a charm. Students quickly recognized the power of additional data: "It's just like a baseball average. The more times someone has been at bat, the more accurate the batting average is likely to be." Students in this class were participating in a project that emphasized "a caring community of learners." By working together to shape class rules and participating in regular class meetings, they had become very skilled at working together, and their teacher consciously avoided external rewards and competition, preferring to have students operate from intrinsic motivation and a personal commitment to everyone's learning.

In a demographically similar classroom just a few miles away, however, the lesson flopped; few students were willing to revise their initial estimates. Very reluctant to admit they were "wrong," students busily defended their initial predictions and refused to use data gathered by other pairs of students, justifying their refusal with criticisms like "you probably chose all your marbles from one part of the bag." The reward system and pervasive sense of competition in this classroom seemed to make it difficult for students to revise their predictions. The very different fates of the same lesson plan in two classes illustrate both the power of classroom social and motivational climate, and the impossibility of judging a "good" lesson on paper.

Figure 12 provides the ingenious schedule developed by Paterson School Number Two, the first US school to integrate lesson study into the school day. Principal Liptak notes the guiding principles behind their scheduling: to maintain high-quality instruction for students and to schedule lesson study in a way that would permit it to become part of the school culture.

In contrast to Paterson School Two, San Mateo lesson study groups generally meet after school (receiving a stipend for after-school time), but conduct research lessons during school hours, using substitute teachers to cover the classes of planning team members who observe the lesson.

**Figure 12**
**It's a Matter of Time:  Scheduling Lesson Study**
**at Paterson, NJ School Two***

With salaries and benefits accounting for over 75 percent of the school budget, there is no question that staff time is the most valuable resource a school has to allocate.  Our decision to implement lesson study, with a school day investment of 80 minutes to two hours each week per participant, was based on the interest and commitment of the teachers and the belief that deep changes in teaching and learning as they occur daily in the classroom are best achieved through collaborative, reflective professional development in the classroom context.

As with any investment there is a downside and a risk factor.  On the downside, when teachers are meeting during the school day, it is in lieu of teaching classes or tutoring students.  There is also risk.  Although lesson study is a highly successful means of professional development and instructional improvement in Japan, it is not easy to adapt the process in an American cultural setting.  Results of adaptations of lesson study in the US setting have yet to be analyzed, and dividends may be deferred.  Our experience these past two and a half years, however, convinced us that lesson study is potentially a powerful vehicle for improving classroom instruction and thus, in our view, well worth the investment of time.

The following principles were considered in developing a lesson study schedule for School Number Two, Paterson, NJ:

- If lesson study is going to become part of the school culture and conducted over a long period with a goal of gradual improvement, then time must be allocated during the school day.  Lesson study has no chance of becoming a prevalent feature of the school culture if it is conducted with a few enthusiastic volunteers working after school.

- Time is one sure measurement of commitment.  When teachers see serious time committed to lesson study, and the administrators taking time to engage in lesson study, they feel confident of a high level of support for the process on a day-to-day basis and over the long haul.

- Lesson study should be scheduled by reallocating currently existing resources.  In our school, it does not rely on "soft" money or the hiring of substitute teachers.

- Quality instruction must be provided in the classroom while the teachers are engaged in lesson study.

Time for lesson study was thus built into the regular school day using non-classroom teachers and pre-service teachers.  School Number Two is a K-8 school in an urban district and thus qualifies for Title I funding.  It has also benefited from the Abbott court decision, which mandated parity in funding for poor school districts.  Funding from these sources has been used to hire English as a second language teachers, reading tutors, and other non-classroom teachers.  The district also provides special area teachers (art, music, physical education, etc.), a guidance counselor, and other non-classroom staff.

*by Lynn Liptak

**Figure 12**
**Continued**

Therefore, it has been possible to pair each classroom teacher in grades 1 to 8 with a non-classroom partner teacher. The partner teacher has contact with the class during the week by teaching during teacher preparation periods, downsizing the class for mathematics or reading, or tutoring individual students. It is the responsibility of the partner teacher to know the students and become familiar with classroom routines. In the event of absence, the partner teacher helps to orient the substitute and assists as needed with the class. The partner teacher often teaches the class while the classroom teacher engages in lesson study. Students in grades 7 and 8 are in special area classes during 80 minutes of the lesson study time.

During our first two years, a group of 16 volunteers met each Monday from 1 p.m. to 3 p.m. to conduct lesson study in mathematics. During the first cycle, it was apparent that the two-hour weekly meeting was only "seed" time. Once we began to collaborate on lessons and test out ideas in the classroom, we did not wait until Monday to continue the process. E-mail communication and discussions before and after school, during lunch periods, and during preparation periods are common. Most importantly, these discussions and observations are focused on how our teaching impacts student learning. We know from research and our own observations that grade level meetings and school management team meetings rarely focus on the lessons that occur daily in the classrooms.

By the end of the second year, a chasm was developing between volunteer lesson study group members and non-participants. To address this, School Number Two decided to go school-wide with lesson study. In August 2001, a three-day lesson study seminar was offered to all teachers. In September 2001, five lesson study groups (kindergarten, grades 1-2, grades 3-4, grades 5-6, and grades 7-8) were launched.

Now all classroom mathematics teachers (with the exception of one who opted not to participate) are scheduled from 80 to 105 minutes per week for lesson study. In grades 7-8, the science teacher also participates, and our first lesson study science lesson is under development. In grades 5-8, special education teachers also participate. The mathematics facilitator attends all meetings and takes minutes, but the leadership of each group is shared among the participants.

While the teachers meet, classes are taught by a combination of partner teachers and special-area teachers. All student teachers attend lesson study with their cooperating teacher (not counted in number of teachers in the figure below which outlines our scheduling for mathematics lesson study).

| Scheduling for School-Wide Lesson Study in Mathematics | | | |
|---|---|---|---|
| Grade Levels | Teachers | Meeting Day, Time | Class Coverage |
| K | 6 | Tuesdays, 1:35–3:00 p.m. | Teaching Assistants |
| 1-2 | 6 | Wednesdays, 1:15–3:00 p.m. | Art, ESL, World Languages, Partner Teachers |
| 3-4 | 5 | Thursdays, 12:55–2:35 p.m. | Music, ESL, Art, Life Skills, Keyboarding, Partner Teachers |
| 5-6 | 6 | Tuesdays & Wednesdays, 10:15–11:00 a.m. | Special Areas: Art, Music, Physical Education, Technology, Life Skills |
| 7-8 | 5 | Mondays, 1:35–3:00 p.m. | Special Areas |

**Figure 12**
**Continued**

Additionally, the grade level teams meet for 40 minutes weekly to deal with issues apart from lesson study. Teachers, by contract, receive one 40-minute preparation period per day. Like most good investments, we expect that the growth and dividends from the time we invest in lesson study will accrue gradually over a long period of time. Improving our teaching in depth is hard, time-consuming work which needs to be done collaboratively and in a supportive setting.

For too long, professional development time has been allocated to outside experts to "train" teachers rather than given to teachers to reflect collaboratively on their practice. We need to tap outside expertise; we need to improve our content and pedagogical knowledge. But the professional development process needs to occur in the context of our classrooms and be driven as an ongoing activity by professional practitioners.

Lesson study — it's about time.

## Intervals between Meetings

In Japan, teachers meet frequently leading up to the research lesson, but take breaks from lesson study at other times of year (for example, when they are busy with other school-wide activities or during holiday season). In the US, teachers may be more likely to have time allotted for lesson study on an unchanging schedule (for example, every Wednesday afternoon). An experienced US teacher gives the following advice about scheduling lesson study:

> *It is important for teachers to understand that lesson study has a definite beginning and ending. Too much time spent in a cycle can be counterproductive.... Teachers should understand that since lesson study is not about making a perfect lesson, a semi-rigid schedule is needed.*[3]

Scheduling research lessons in advance for particular dates throughout the year — rather than waiting until you feel completely "ready" — can speed up your progress, much the way a deadline can catalyze work on other projects. Likewise, scheduling in advance the second teaching of the lesson can streamline the lesson revision process. The research lesson needs to be discussed soon after it is held (preferably on the same day). Intervals between other lesson study events need to be given some thought. In Japan, re-teaching generally occurs soon after the original lesson (within a few days or a week), and some US schools follow this practice. For example, the visitors to the April 2002 public research lessons in San Mateo-Foster City School District (California) were able to see lessons taught, revised, and then taught to a different class two days later. Many of us were amazed at the dramatic differences between the first and second teachings, apparently produced by minor modifications in the lessons, such as asking groups to come up with additional solutions after they had solved the problem in one way.

Other US schools choose a longer interval before re-teaching. The schedule at Paterson Public School Number Two generally allows about three weeks between the first and second teachings of a lesson; Bill Jackson explains that "it is sufficient time to do the revision...and it forces us to do the revision quickly. We felt we spent too much time between first and second teachings in previous years."[4] This schedule requires the second teacher to hold back on teaching a particular topic, but this has worked well at their site "because it allows that teacher to throw in a few extra lessons to prepare for the final study lesson. One problem we find in study lessons is students' lack of prior knowledge. An extra week or two to firm up some basic concepts helps."[5] Like the teachers at Paterson Public School Number Two, you will want to adjust your lesson study cycle to reflect what you learn the first time around. The sooner you do an actual research lesson, the sooner you will be able to learn from the process.

## Lesson Study: One More Burden for Teachers?

Prior sections of this chapter describe the basic activities of lesson study and the time required for each, explore scheduling options, and contrast the strengths of video and live observation. Far more challenging than the logistics of scheduling is the know-how needed to build an effective lesson study effort. The next chapter, a step-by-step guide to lesson study, provides practical guidance on how to build lesson study in your setting.

# Chapter 5 Notes

[1] Nick Timpone, Teacher, Paterson Public School Number Two, Questionnaire Response, January, 2001.

[2] Developmental Studies Center (1994). *At home in our schools*. Oakland, CA: Developmental Studies Center.

[3] Bill Jackson, Teacher, Paterson Public School Number Two, Email, January 1, 2002.

[4] Ibid.

[5] Ibid.

# 6

# Pioneering Lesson Study in Your School: A Step-by-Step Guide

> **My lesson planning has changed this year because of lesson study. I'm moving away from "What are the activities I'm doing?" to "What is it that I want kids to get?"**
>
> - Jacqueline Hurd, Teacher[1]

Earlier chapters focused on the big ideas behind lesson study. This chapter focuses on the practical details of lesson study. Following the step-by-step guide that is summarized in Figure 13, this chapter lays out lesson study's steps: forming a group; focusing the group's work; planning the research lesson; conducting, discussing, and revising the lesson; and reflecting and planning next steps.

## Step 1. Form a Lesson Study Group

There are four major activities in forming a lesson study group: (1) recruit members; (2) make a specific time commitment; (3) set a schedule of meetings; and (4) agree on ground rules for your group.

### Recruit Members

Figure 14 suggests some strategies for building a lesson study group. Perhaps, like the San Mateo teachers, you will distribute a one-page flyer on lesson study and invite interested volunteers to an informational meeting. Or perhaps an existing group in your setting (like the Mathematics Study Group at Paterson School Two) provides a natural start for lesson study.

Offering lesson study as an alternative way to meet an existing obligation recognizes that many teachers need something taken off their plates before they have room for something new. Groups involved in activities such as implementation of a new curriculum or standards, improvement of instruction in a particular area, long-term planning, or program quality review might find lesson study an effective, classroom-focused means to pursue their goals. Or perhaps you want to form a group initially with just a few trusted colleagues who also enjoy the challenges of trying an emerging innovation. Whatever approach you choose, remember to be open and welcoming to curious outsiders.

**Figure 13**
**Lesson Study Steps**

**1. Form a Lesson Study Group (1-2 meetings).**

- Recruit members for a lesson study group (see Figure 14 for tips).
- Make a specific time commitment.
- Set a schedule of meetings.
- Agree on ground rules for your group.

**2. Focus the Lesson Study (1-4 meetings).**

- Agree on a research theme (or main aim) that captures your long-term goals for students.
- Choose a subject area (e.g., science, social studies).
- Choose a unit and lesson, and agree on goals.

**3. Plan the Research Lesson (3-6 meetings).**

- Study existing lessons.
- Develop a plan to guide learning, following the examples in the appendices. The plan to guide learning provides practical and conceptual guidance for both instructor and observers, and addresses long-term goals for students as well as specific goals for the lesson and unit. It includes a data collection plan.
- Consider an outside specialist.

**4. Teach and Observe the Research Lesson (1 lesson).**

- Collect data as planned.

**5. Discuss and Analyze the Research Lesson (1 meeting following research lesson on same day, additional meetings if desired).**

- Follow a carefully-structured agenda.
- Focus discussion on the data collected at the research lesson.
- Consider how to revise the lesson/unit/instructional approach.

**6. Reflect and Plan the Next Steps (1-2 meetings).**

- Re-teach the revised lesson, if desired, repeating steps 4 and 5 one or more times.
- Discuss the benefits and challenges of your lesson study effort, and what you would like to change in the next cycle.
- US teachers who choose to participate in lesson study are pioneers. Celebrate!

## Figure 14
## Strategies for Building a Lesson Study Group

**Strategy: Build Awareness, Recruit Volunteers.**

**Resources:** Invite colleagues to read an article or watch a videotape on lesson study; identify interested collaborators.

**Strategy: Transform an Existing Group.**

**Resources:** Existing groups in your district or region may provide a natural venue for lesson study:

- *Committees on curriculum, standards, assessment.* Research lessons can bring their ideas to life in the classroom for others to see.
- *Mentor teachers, coaches, subject matter specialists.* Research lessons provide a way to refine and spread their ideas about good practice, perhaps by devoting a professional development day to research lessons.

**Strategy: Reshape Current Work to Include Lesson Study.**

**Resources:**

- *Grant-funded work.* Perhaps a funder would welcome an open-house research lesson instead of the traditional final report. Research lessons, because they will be observed by other teachers, provide built-in accountability and can have ripples of influence that naturally disseminate grant-funded work.
- *Professional development credits.* Rather than steer teachers toward individual coursework or one-shot conferences, how about making lesson study groups an option?
- *Program quality review, school improvement plans.* Lesson study can provide the structure for setting goals, improving instruction, and assessing student development.
- *Routine performance reviews.* Tenured teachers might be given the option of conducting a research lesson in lieu of current requirements (such as observation by the principal).

**Strategy: Contact Local Members of a Local Union or Subject-Matter Organization.**

**Resources:** Teachers unions and subject-matter associations (such as National Council of Teachers of Mathematics) have been leaders in publicizing lesson study. Science museums, schools of education, and other local institutions may help you network to build a lesson study group.

**Strategy: Find a School with a Supportive Mission.**

**Resources:** An existing or planned magnet school or professional development school could incorporate lesson study as a key operating principle.

**Strategy: Gather Your Buddies.**

**Resources:** Working with a few colleagues, start a small lesson study group, where you can learn how to adapt lesson study to your setting. If you can make lesson study useful to your group, support and interest are likely to follow.

**Strategy: Look Online.**

**Resources:** Information about lesson study groups is beginning to be available online. Live observation of lessons seems integral to lesson study now, but may become less so as video technology advances.

## Make a Specific Time Commitment

Is lesson study worthwhile? Can members of your group make a commitment of time to make it happen? Perhaps more than any other kind of professional development, lesson study depends upon teachers to bring it to life. Japanese lesson study groups probably meet one to four times per month on average, but they may meet frequently leading up to a research lesson, and not meet at all during busy times of the year. If there is a summer school or year-round school in your area, you might want to consider a summertime lesson study workshop, so that teachers can explore lesson study free from the demands of the regular school year. (The down side to summer lesson study is that it's hard to know the students as well as you would know your own class during the school year.)

What contribution will you expect from group members? Some groups form with the understanding that not all members want to teach a research lesson, and that no one will be pressured to do so. Others expect all members will take a turn. It makes sense to discuss these expectations up front.

## Set a Schedule of Meetings

Schools are busy. Japanese teachers often block out times for lesson study meetings in advance for the entire school year. This ensures that the various parts of lesson study (for example, faculty meetings, small group meetings, and research lessons) will all occur in the right sequence, and that necessary preparations (e.g., substitutes) can be arranged. Wouldn't it give your group a healthy feeling of stability to know that each lesson study group in your school or district has committed to teach research lessons on particular dates during the next year? Researcher Makoto Yoshida recommends a schedule in which each lesson is taught two or three times in different classes, and discussed and revised after each teaching. (Figure 10 in Chapter 5 presents his thoughts on why.) The public research lessons taught at Paterson School Number Two in 2000 had already been taught, observed, and refined once before the lesson study open house.

## Agree on Ground Rules for Your Group

Admit it. The thought of having colleagues in your classroom to observe a lesson is scary. (Japanese teachers also think it's a scary, but essential part of being a real teacher.) While the protocol for lesson observation and debriefing introduced later in this chapter can ease these tasks, lesson study groups may also want to make explicit norms or ground rules for their work together. For example, you may want to discuss in advance how decisions will be made, responsibility shared, time used, and suggestions offered. Rotating the job of facilitator and saving a few minutes at the end of each meeting to reflect on the meeting's strengths and shortcomings can help build a group process that works for all members.

Effective lesson study groups often seem to have three characteristics:

- **Egalitarian discussion.** Lesson study differs from mentoring or coaching in its emphasis on lesson inquiry conducted by equals. Can you develop operating procedures that assume that every teacher, whatever his or her level of experience, has something valuable to contribute to the study of student learning and development?

- **Shared ownership and responsibility.** As Chapter 3 highlights, the purpose of lesson study differs in various settings. But even when its primary purpose is to develop curriculum or enable an individual practitioner to reflect on and improve practice, there is a strong sense of collaborative support. One US group promotes the sense that the lessons are "our" lessons by planning together without knowing who will teach the lesson, and then selecting the instructor's name from a hat near the end of the planning process. Japanese teachers share many school-wide activities (such as the sports festival, arts festival, school trips, and so forth) that naturally forge a sense of shared responsibility for students and few competitive traditions that pit teachers against one another. The point of lesson study is not to polish the skills of a few star teachers but to help all teachers progress, in order to reach as many students as possible with successful lessons and a coherent experience.

- **Emphasis on students, not teacher.** Lesson study focuses on student learning and development. A US teacher pointed out the difference from lesson observation familiar to US teachers: "In the US, if you are being observed, it's a critique of you. Lesson study focuses on student learning, on student 'aha's.' It takes what we're doing to a more professional level."[2] Letting data speak for themselves is often more powerful than evaluative statements. For example, "50 percent of the students raised their hands in response to your questions" is preferable to "I thought your questions succeeded (failed) in getting students to raise their hands."

Your group may want to begin norm-setting by looking at ground rules developed by others. I liked the following ground rules that I saw on a bulletin board at a US middle school:

- Communicate clearly and listen carefully.

- Respect the views of others.

- Share your views willingly.

- Ask and welcome questions for clarification.

- Be open to the ideas and views presented.

- Honor time limits.

- Stay on task.[3]

## Step 2. Focus the Lesson Study

Three activities are suggested to focus your lesson study: (1) agree on a research theme, (2) choose a subject area, and (3) select a particular unit and lesson. Each is described in this section.

## Agree on a Research Theme That Captures Your Long-Term Goals for Students

Like the teachers at Komae School, introduced in Chapter 1, you may want to begin your lesson study with the following two questions:

- Ideally, what qualities will students have when they graduate from our school?

- What are the actual qualities of our students now?

By comparing the "ideal" and "actual" qualities of students, you can locate the most meaningful focus for your group's lesson study. Figure 15 provides a handout to facilitate this discussion; pose each question and allow teachers to spend some time jotting down their ideas about ideal, actual, and the gap between the two. A Japanese teacher defined lesson study as follows: "The students' actual situation right now is the starting point for your journey and students' ideal qualities are your destination. Lesson study is the road that links the two."[4]

Perhaps you are interested in lesson study primarily as a way to improve mathematics instruction now, not in students' qualities five years from now. Even so, you may find it worthwhile to think about the qualities you would like your students to have five years from now. You'll probably find that mathematics (and every other subject you teach) is inseparable from your broader goals for students. Qualities like persistence, reflection, and responsibility for learning are essential to learning mathematics, and, conversely, mathematics is one important arena in which students learn (or fail to learn) these qualities. US researchers were surprised to find that Japanese teachers began mathematics lesson study with the question "What kind of children do we want to raise in this school?"[5]

By reflecting on the gap between the ideal and the actual, choose your "research theme" (also called a "research focus" or "main aim" of lesson study). As the examples included in Chapter 4 illustrate, the research theme is usually a broad goal that is compelling to teachers from all grade levels and many points of view, such as to build desire to learn, responsibility and initiative as learners, and understanding and facility with subject matter. Is your research theme something important, enduring, and fundamental to your mission? Is it the sort of goal that brought you into education?

## Choose a Subject Area

If you haven't yet chosen the subject area (such as mathematics) for your lesson study, your group may want to think about the following questions:

- What subjects or parts of school life are the most difficult for students?

- What subject do teachers find most difficult to teach?

- In what subjects are there new curricula, frameworks, or standards that teachers want to understand and master?

**Figure 15**
**Choosing a Research Theme (Main Aim) for Lesson Study**

Think about the students you serve.

**<u>Your Ideals:</u>**

Ideally, what qualities would you like these students to have five years from now (or alternately, when they graduate your institution)?

**<u>The Actual:</u>**

List their qualities now.

**<u>The Gap:</u>**

Compare the ideal and the actual. What are the gaps that you would most like to work on?

**<u>The Research Theme:</u>** (The goal, research focus, or main aim of lesson study)

By comparing the ideal and actual student qualities, select a focus for your lesson study. State *positively* the ideal student qualities you choose to work on. For example, teachers in a Japanese school serving many low-income and minority students chose the following goal:

> For students to develop fundamental academic skills that will guarantee their advancement, and a rich sensibility about human rights.

Your research theme:

Japanese teachers use lesson study not just to improve academic instruction, but to work on other areas as well: for example, to improve class meetings, whole-school activities, school climate, and integration of second-language learners.

Once you have chosen a research theme (main aim) for your lesson study and chosen the subject area, it's time to think about how you will work toward your goals. What kinds of instruction (and other elements of school life) will enable you to nurture the "ideal students" you imagined? What instructional changes are needed? What evidence from student work and behavior would show that students are moving toward these goals?

You may find it helpful to lay out your thinking in a research map that shows the connections between your long-term goals for students and your planned research lesson(s). Figure 16 provides the research map from Komae School Number Seven and shows how their research theme (in the dotted box) selects certain qualities from the ideal student profile that will be the focus of research lessons. Reading from the bottom of the figure, the research map shows what strategies will be used during research lessons and what evidence will be gathered ("methods and measures"), why instruction will be designed in this way ("research hypotheses"), and what the lower-, middle- and upper-grade teachers expect to see in their students if progress is being made toward the school-wide research theme. This research map was developed collaboratively by the teachers at Komae School, and provides a framework for their research lessons throughout the year, including *Can You Lift 100 Kilograms?*[6]

The lowest box in the research map shows that teachers plan to look at four different aspects of instruction: curriculum, learning materials, learning activities, and strategies for teaching and evaluation. (These elements are addressed in detail in each research lesson plan.) The next box up shows the teachers' hypotheses that students will deepen their perspectives and ways of thinking during lessons in which they take initiative and learn eagerly. Looking at the central rectangles, we see that teachers expect enjoyment of science to show up in increasingly sophisticated ways from the lower grades ("participate happily in learning") to the middle grades ("eagerly use five senses") to the upper grades ("get pleasure from solving problems").

A blank version of the Komae research map is provided in Appendix 6. It may help you to create a research map that captures the actual and ideal qualities of your students, defines your hypotheses about the kinds of instruction that will foster those ideal qualities, and lays out the qualities you would expect to see at each age level if students are developing toward the ideal.

The lesson study research map can be particularly useful in whole-school lesson study, because teachers from different grade levels can use it to relate their school-wide research theme to students of different ages. In school-based lesson study, often the research map is developed in a back and forth process between the grade-level groups and the whole faculty.

## Figure 16
## Komae School Research Map

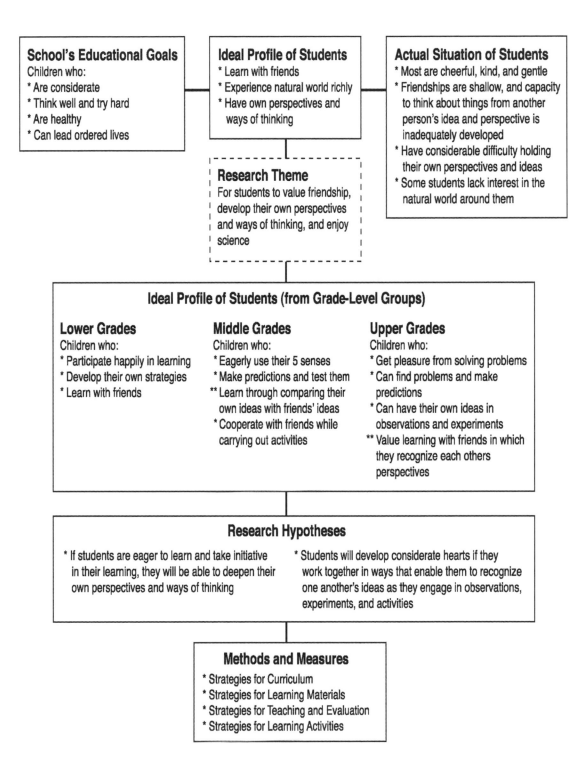

**School's Educational Goals**
Children who:
* Are considerate
* Think well and try hard
* Are healthy
* Can lead ordered lives

**Ideal Profile of Students**
* Learn with friends
* Experience natural world richly
* Have own perspectives and ways of thinking

**Actual Situation of Students**
* Most are cheerful, kind, and gentle
* Friendships are shallow, and capacity to think about things from another person's idea and perspective is inadequately developed
* Have considerable difficulty holding their own perspectives and ideas
* Some students lack interest in the natural world around them

**Research Theme**
For students to value friendship, develop their own perspectives and ways of thinking, and enjoy science

**Ideal Profile of Students (from Grade-Level Groups)**

**Lower Grades**
Children who:
* Participate happily in learning
* Develop their own strategies
* Learn with friends

**Middle Grades**
Children who:
* Eagerly use their 5 senses
* Make predictions and test them
** Learn through comparing their own ideas with friends' ideas
* Cooperate with friends while carrying out activities

**Upper Grades**
Children who:
* Get pleasure from solving problems
* Can find problems and make predictions
* Can have their own ideas in observations and experiments
** Value learning with friends in which they recognize each others perspectives

**Research Hypotheses**

* If students are eager to learn and take initiative in their learning, they will be able to deepen their own perspectives and ways of thinking

* Students will develop considerate hearts if they work together in ways that enable them to recognize one another's ideas as they engage in observations, experiments, and activities

**Methods and Measures**
* Strategies for Curriculum
* Strategies for Learning Materials
* Strategies for Teaching and Evaluation
* Strategies for Learning Activities

## Choose a Unit and Lesson, and Agree on Goals

Once you have agreed on the research theme and subject area for lesson study, it's time for the group of four to six teachers who will plan the research lesson to choose its topic. You may wish to choose a topic that is:

- fundamental to subsequent learning
- persistently difficult for students or disliked by them
- difficult to teach or disliked by teachers
- *new to the curriculum.*

You may also choose whatever topic happens to be taught at the time of the scheduled research lesson, with the idea that all lessons should provide a window on your long-term goals and your approach to this subject.

US teachers report that the task of choosing a unit and lesson for the focus of lesson study can be very easy or very difficult. Some lesson study groups report that diagnostic tests clearly identify a problem (for example, student difficulty with word problems) but may not provide a clear idea about the kinds of lessons that would better build student understanding. One lower-grades lesson study group originally thought their students were having difficulty with the place value of ones and tens, but as they began to design a lesson they found that the real problem seemed to be students' fluency with combinations of numbers that make twenty. Another group changed their lesson topic after several meetings; as one member reported, "It takes time coming to one mind. You fumble along the way. You spend time going down dead ends. At our third meeting we realized we were wrestling with a topic we didn't care enough about. You have to be patient with the process."[7]

Once you've chosen a topic — let's say, introduction of fractions — you need to specify the goals for that topic. In Japan, the goals for a topic are often taken directly from the national *Course of Study*, and are quite straightforward. Examples include to know how to calculate the area of a rectangle, or to know three properties of magnets. These goals become the unit goals (see instructional plans in Appendices 2 and 3), and lessons are planned that provide building blocks toward the unit goals. For example, in the video *Can You Lift 100 Kilograms?*[8] one lesson goal was for students to discover how hard it is to lift a heavy object by hand and to actively consider what things can be used to ease lifting. This goal, the plan explains, will set the stage for students to actively pursue "as their own problem" the knowledge about levers that is to be developed during the unit.

Along with specific goals for subject matter, research lessons simultaneously focus on broad, long-term goals for each subject area and for student development, which are also laid out in the *Course of Study*: for example, to learn eagerly, take initiative, develop scientific habits of mind, love nature, be active problem-solvers, and notice mathematics in daily life. The simultaneous focus on specific and broad goals makes lesson study confusing to newcomers. People sometimes ask me in exasperation, "Which is lesson study *really* focused on, teaching of specific topics like addition of fractions or long-term goals like love of learning?" The answer is both. This makes sense if you consider that broad, long-term goals like responsibility and love of learning are built up through daily lessons, and that, in turn, these basic student qualities greatly support or constrain how teachers can teach daily lessons.

To plan the research lesson, you need goals at four levels:

- goals specific to the lesson
- goals specific to the unit
- broad goals of the subject area
- long-term goals for student development.

Figure 17 provides an example of goals at each of the four levels.

Once all four levels of goals are specified, it is time to plan a research lesson and the larger unit of which it is a part. But don't feel that your goals are cast in stone. As you immerse yourself in study of existing lessons and in observation of students, you may refine your goals for the lesson and unit, and perhaps even rethink your long-term goals for students.

## Figure 17
## Examples of Four Levels of Lesson Study Goals

### Level 1: Goals Specific to the Lesson

- Be motivated to find out the principles of levers in subsequent lessons.
- Find out what businesses and institutions are in the neighborhood of the school.
- Discover that the circumference of a circle is always about three times its diameter.

### Level 2: Goals Specific to the Unit

- Understand that the force needed to lift an object of constant weight with a lever changes depending on the location of the object and force.
- Develop an awareness of the local community and one's role in it.
- Understand how to calculate the area of a circle and how the area of a circle relates to the area of a rectangle.

### Level 3: Broad Subject-Matter Goals

- Develop scientific ways of thinking such as use of the five senses, use of evidence to warrant assertions, and use of controlled investigation.
- Develop a foundation for the qualities of citizenship needed to be members of a democratic, peaceful society.
- Actively use prior knowledge to solve novel mathematics problems.

### Level 4: Long-Term Goals for Student Development

- Take initiative as learners.
- Learn with desire.
- Value friendship.
- Work cooperatively with others.

## Step 3. Plan the Research Lesson

Japanese teachers spend more time planning a research lesson than they would typically spend planning a regular daily lesson. Two key elements of the planning are discussed here: the study of existing lessons; and the development of a plan to guide learning that records the group's thinking and later serves as a guide for teaching, observing, and discussing the lesson. These processes are interwoven and inform each other.

### Study Existing Lessons

Lesson study is most productive when educators build on the best existing lessons or approaches, rather than reinventing the wheel. Just as you might want to develop your artistic or musical taste by immersing yourself in many works of art and music, try to immerse yourself in others' lessons through whatever means you can. When asked how he could help his faculty learn to teach a new subject, Life Environment Studies, a Japanese principal said:

> *The way to improve Life Environment Studies is to see many good actual examples. We can do that by going to lots of schools that are doing presentations and research lessons on Life Environment Studies. Many people from this school have gone. Each school has its own way of approaching the new subject. Some are appropriate for your school, some aren't. What works elsewhere might not work at your school because the children are different. So you need to see lots of examples.[9]*

Japanese teachers gather lesson examples from many sources, including textbooks, research lessons at other schools, books and videotapes published by teachers, and even (I was once surprised to find) demonstration lessons observed in the US. In addition to published volumes, unpublished reports and videos of lesson study are available through many organizations, including teacher centers, professional organizations (such as the science teachers' association) and individual schools. Japanese teachers can also attend live open house research lessons conducted by districts, professional societies, the 73 national elementary schools, and schools that have received special research funds. Teachers in Hiroshima studied by Makoto Yoshida saw about ten research lessons every year.[10]

US teachers who pioneer lesson study will have a tougher time locating resources. In Appendix 7 we list some resources, such as the TIMSS videotapes. Inquiries to university-based educators, museums, regional education laboratories, projects funded by the National Science Foundation, and professional organizations are also likely to uncover curricula that can provide an excellent starting point for lesson study. You will also want to search out other resources, such as the local teacher who has an excellent reputation, mentor teachers, and teachers whose work is highlighted in professional publications. Don't underestimate the importance of seeing many models related to the practice you are trying to forge. A Japanese educator reflected on the importance of borrowing from others:

> *If you shoot for originality too early in your development as a teacher, you're likely to fail. Initially, you must take a lot from others. But ultimately, to move to a higher level of teaching, your lesson must become your own original thing, not simply imitation of others. But it's through imitating others' lessons you create your own authentic way of teaching.*[11]

If your group searches out and studies the best existing lessons, it will result in a better research lesson and help create a system that learns rather than one in which every group of educators reinvents the wheel. As Isaac Newton reflected on the cumulative progress of science, "If I have seen further...it is by standing on the shoulders of giants."[12]

## Develop a Plan to Guide Learning

The plan to guide learning (*Gakushuu Shidouan*) guides the teaching, observation, and discussion of the research lesson, and captures the inquiry that occurs during lesson study. Appendices 2 through 4 provide several examples of these instructional plans, which have a much broader function than the lesson plans used by Japanese teachers for daily lessons. Appendix 5 provides a blank template for your plan. Because the plan to guide learning is complex, it may be helpful to think about its elements in three concentric circles, with the research lesson plan at the center, the unit plan in the next ring out, and the whole instructional plan as the outermost ring, as illustrated in Figure 18.

**Figure 18**
**The Three Concentric Circles of the Plan to Guide Learning**

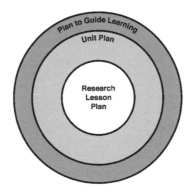

### The Innermost Core: The Research Lesson Plan

Located near the middle or end of the instructional plan is a detailed plan for the research lesson itself, often called "today's lesson." (See "Today's lesson," section 7 in the levers lesson plan found in Appendix 2, and "Lesson plan for this hour," section 8 in the mathematics lesson plan found in Appendix 3.) The research lesson plan answers the central question:

What changes in student thinking will occur during the lesson, and what will provoke them? This research lesson plan is usually written out in three or four parallel columns that contain:

- the questions, problems, and activities to be posed by the teacher
- the anticipated student responses
- the teacher's planned responses to the students
- points to notice during the lesson (or "evaluation").

The following list of questions may help to guide the planning of your research lesson:

1. What do students currently understand about this topic?

2. What do we want them to understand at the end of the lesson?

3. What is the "drama," or sequence of questions and experiences that will propel students from their initial understanding to the desired understanding?

4. How will students respond to the questions and activities in the lesson? What problems and misconceptions will arise? How will the teacher use these ideas and misconceptions to advance the lesson?

5. What will make this lesson motivating and meaningful to students?

6. What evidence about student learning, motivation, and behavior should be gathered in order to discuss the lesson and our larger research theme? What data collection forms are needed to do this?

As the list suggests, research lesson planning is different from the lesson planning familiar to US teachers. Teacher Heather Crawford describes the planning of a research lesson:

> It is challenging — to try and think about the students' solutions to the problem before they do it, and to try and get all of the answers they might come up with. You have to think about things from the student's point of view and that is a big change. [13]

She contrasts research lesson planning with the lesson planning that previously occurred at her school:

> ...[now] we think a lot more about the motivation for the lesson and making sure that the kids have the prior knowledge that they need before we teach each lesson.... Before we did lesson study we really didn't think about what the student responses would be to the questions. When we posed a problem we never really thought about what the kids would come up with. It was..."Well, we hope they get the right answer and if they don't then we will deal with it." Now we are really thinking about, "Well, what if this answer were to come up? How would we deal with it?" [14]

A Japanese teacher writes that he builds the research lesson plan by imagining: "If the teacher says this, how are the children likely to respond?" He goes on to say that "the ability to imagine how students will respond in an actual lesson is one strength of a good teacher."[15] In a similar vein, another Japanese teacher describes how he first "tests" his research lesson plan on an imaginary audience of children:

*Something that you often hear said is "The research lesson was constrained by the lesson plan, so it didn't go well." I think we teachers must do a lesson twice. The first time is with imaginary students as partners, and the second time is with the actual students of the classroom as partners. When you feel constrained by the lesson plan, it's because the first lesson with imaginary students failed. If you put your hopes into a lesson that is based only on your thinking, without considering the ideas it will elicit from students, you will face a painful gap with reality.[16]*

You might want to flesh out your research lesson plan by having one member of your group teach it and the others serve as students, imagining what responses the actual students will come up with. In addition to the imagined storyline that will take students from their initial understanding to a new understanding at the end of the lesson, the research lesson plan contains other information useful to team members and outside observers. The column of "points to notice" alerts observers about what to look for at each stage of the lesson. For example, observers might be cued to notice whether students are eager to investigate a problem introduced by the teacher, what methods they use to compare the size of two rectangles, or how their drawings of hot air change after conducting an experiment.

Additional columns or sections may note the time allocated to each element of the lesson and the materials needed for it. Finally, the lesson's expected outcomes are summarized in a list of lesson goals or objectives.

## Second Circle: The Unit Plan

Moving out from the immediate research lesson plan, we find information on the larger unit of which the research lesson is part. This information includes the unit objectives and unit instructional plan (items 2 and 6 in Appendix 2). While US educators often think of lesson study as focusing on a single lesson, in fact the whole unit is under study, although just one lesson is typically observed. The unit plan shows how the observed research lesson fits into a series of lessons. From the unit plan, observers can tell, for example, whether the primary function of the lesson is to motivate subsequent study of the topic, to teach a particular concept, or to help students consolidate and apply what they have learned in prior lessons. Sometimes the instructional plan provides information even beyond the unit. For example, it may show how the research lesson topic connects with material taught in prior or subsequent years of schooling (see the mathematics lesson plan provided in Appendix 3).

## Outer Circle: The Research Theme

Finally, several items in the plan to guide learning explain the connection between the lesson and the research theme. (See, for example, "connection with research theme," in Appendix 2). These items help observers understand the educational philosophy behind the research lesson, and how it connects with the planners' long-term hopes for students. For example, the planners of the levers lesson proposed the hypothesis that "if children can discover that even in the commonplace tools of everyday life the rules and laws of nature are at work, it should be a joyful experience for them."[17]

**Data Collection Plan**

An element of the plan to guide learning that cuts across all three circles is the plan for data collection. As noted, a column of the research lesson plan labeled "points to notice" or "evaluation" typically guides observers to notice particular aspects of the lesson. Although guests from outside the school may not be asked to collect data, members of the lesson study group and school faculty are usually given specific data collection assignments and forms to aid in data collection, which might include a seating chart, list of members of each student group, records of students' prior thinking, checklists for noting features of student work, forms for recording the participation of each member of a small group, or forms to collect other data the lesson study group deems most relevant to their goals.

The particular data gathered vary greatly, depending on the research theme. For example, teachers at Komae School Number Seven studied how students' ideas about levers changed, whether the normally quiet students participated in each small group's discussion, and whether there was evidence of student engagement, such as "shining eyes" and *tsubuyaki* (under-breath exclamations, or aha's). Data collected during lesson study typically include evidence of academic learning, motivation, and social climate. Figure 7 in Chapter 4 provides examples of the types of data collected during research lessons. Although data collection typically focuses heavily on students, it is also common to make a record of the teacher's speech and of the time devoted to each element of the lesson. In this way, teachers can later analyze issues of interest to them (for example, how they allocated time, and their use of non-evaluative responses like "I see").

**The Uses of a Plan to Guide Learning**

In summary, the plan to guide learning represents the thinking of the whole lesson study group about three concentric layers of practice: the lesson itself, the larger unit and subject area of which it is part, and the even larger domain of students' long-term development. As you move from planning to doing the research lesson, the plan will serve several purposes:

- Support the research lesson instructor, by providing a detailed outline of the lesson and its logistical details (such as time, materials).

- Guide observers, by specifying the "points to notice" and providing appropriate data collection forms and copies of student handouts.

- Help observers understand the rationale for the research lesson, including the lesson's connection to goals for subject matter and students, and the reasons for particular pedagogical choices.

- Record your group's thinking and planning to date, so that you can later revisit them and share them with others.

Because the plan to guide learning plays several important roles and because it is so unfamiliar to Americans, your group might want to analyze carefully the plans in Appendices 2 through 4. They provide a good window on the planning that leads up to the research lesson and a framework for your own plan, which can be created using the template in Appendix 5. Note each element in the plan and allot time to discuss each, including the ideal and actual situation of the students, long-term goals for students, lesson and unit goals, learning flow of the entire unit, detailed lesson plan, and so forth.

Using a Japanese-style plan to think through the research lesson itself as well as the larger unit may reveal unexpected benefits. A Japanese teacher writes that the plan to guide learning is:

> *...a hypothesis for the lesson.... The instructional plan needs to express the problem that necessitated this lesson, what the lesson newly proposes, and it must include the teacher's own vision of education, of children, of mathematics. ...it's a great deal of work. But through writing it, you become aware of how you think about lessons and about mathematics.*[18]

Another Japanese teacher underlines the value of writing the instructional plan:

> *By writing, you can organize your own ideas. Writing a lesson plan is important for that reason alone. That's why we ask our students to write. By the same token, that's why it's important for teachers to write. By writing, we can organize our thinking about the goals of the lesson, the content, the methods.*[19]

### Consider an Outside Specialist

Another element that may enhance your lesson study is an outside specialist — a teacher or researcher who is knowledgeable about the subject matter under study, how to teach it, or both. In fact, it is often most effective to involve an outside specialist early on in your group's work, so that the specialist has a chance to help shape the lesson, to suggest curricular resources, and eventually to serve as a commentator on your research lesson. Lesson study groups at both Paterson School Number Two and San Mateo-Foster City have drawn on the expertise of a number of researchers in mathematics, mathematics education, lesson study, Japanese education, and children's learning.

Figure 6 (Chapter 4) highlights the role of the outside specialist in Japanese lesson study. Though not required, the specialist may play a crucial role within a school and beyond. In Japan, well-known commentators may visit dozens of schools every year and play a role in disseminating lessons and approaches from other schools. If you choose to use an outside specialist, make sure he or she understands the collaborative, student-focused nature of lesson study. You may want to share Figure 3 (Chapter 2) with the specialist, in order to highlight the differences between lesson study and traditional expert-led professional development. In lesson study, the role of the outside specialist is to raise questions, add new perspectives, and be a co-researcher — not to give advice.

### Step 4. Teach and Observe the Research Lesson

By the day of the research lesson, your group will have developed a plan to guide learning that:

- Captures your best collective thinking about how to teach this particular topic to these students.

- Explains your long-term goals for students and how you will bring them to life in the classroom.

- Lays out the problems and activities to be posed by the teacher, and anticipates the responses of students.

- Anticipates problems that may occur and how they will be handled (for example, how calculation errors will be handled).

- Lists practical information, such as the materials needed and how much time will be allocated to each part of the lesson.

- Tells observers what to look for during each part of the lesson and what data to collect, and provides needed forms (for example a student seating chart, prior work from each "focus child" of interest, or note-taking forms specifically designed to collect data of interest).

Now it's time for the most interesting part of any research effort — seeing how your ideas fare in practice. The data collected by the observers will enable you to slow down the "swiftly flowing river" of instruction in order to study it. Your group should decide in advance what data are to be collected (see Figure 7 in Chapter 4 for examples), and assign particular individuals to each data collection task. In Japan, research lessons are often documented by audiotape, videotape, still photography, student work, and narrative observation notes. If you can assign two or more teachers to collect data on a particular child or group of children, you create a natural opportunity for teachers to learn about the strengths and shortcomings of their own observational skills. For example, a Japanese teacher told me that she had learned to notice children's nonverbal behavior after she compared her own observational notes with those of another teacher and realized that she tended to miss nonverbal cues.

Figure 19 provides lesson observation guidelines; they deserve careful review by each member of your team and each observer of the research lesson. The role of the observers during lesson study is to collect data. As US researchers have noted, the observers are supposed to function as "an extra set of eyes, not an extra set of hands."[20] When observers help students, it's difficult to draw inferences about how well the lesson worked. It's important to let students know in advance that the extra teachers in the room will be studying the lesson, not helping the students. That way, students won't think they've encountered a roomful of exceptionally unhelpful adults!

Given the many logistics involved in conducting a research lesson, you may want to divvy them up among group members, designating different members to:

- Obtain needed materials for the lesson.

- Copy the instructional plan for observers.

- Take notes at the post-lesson discussion.

- Facilitate the post-lesson discussion.

- Keep the rest of the school informed.

**Figure 19**
**Protocol for Observation and Discussion of a Research Lesson**

---

**OBSERVATION OF RESEARCH LESSON**

1. Do not help students or otherwise interfere with the natural flow of the lesson.

2. Collect data as requested in advance by the research lesson planning team, or focus your observation on the "points to notice" laid out in their instructional plan.

---

**DISCUSSION OF RESEARCH LESSON**

1. **The Instructor's Reflections.** The instructor describes her or his aims for today's lesson, comments on what went well and on any difficulties, and reflects on what was learned in planning and conducting today's lesson (5 minutes or less).

2. **Background Information from the Lesson Study Group Members.** The lesson study team members explain their goals for students (both lesson goals and long-term goals) and why they designed the lesson (and unit) as they did. They describe changes made to the lesson design over time.

3. **Presentation and Discussion of Data from the Research Lesson.** Lesson study team members (followed by observers, if any) present and discuss data on student learning, engagement, and behavior from the research lesson and the larger unit of which it is a part. The data may include student work, a record of questions by the teacher and/or students, narrative records of all activities by particular children, record of the blackboard, etc., that have been agreed upon in advance. What do the data suggest about the students' progress on the lesson goals and goals for long-term development?

4. **General Discussion.** A brief free discussion period, facilitated by a moderator, may be provided. The focus is on student learning and development, and on how specific elements of lesson design promoted these. The moderator may elicit and group comments, or designate particular themes for discussion, so that there is ordered discussed of key issues, rather than a "point-volleying session" (an apt phrase from the Lesson Study Research Group's protocol, see table note). Comments of a sensitive nature may be conveyed privately at a later time.

5. **Outside Commentator** (optional). An invited outside commentator may discuss the lesson.

6. **Thanks.** Particularly if the gathering is large, it is common for an administrator to thank the instructor, planners, and attendees for their work to improve instruction. In addition, participants usually begin their comments by thanking the lesson instructor and mentioning something they learned from watching the lesson.

---

Note: This protocol is based on the discussion agenda from *Can You Lift 100 Kilograms?* (www.lessonresearch.net), and on the more detailed protocols available at www.globaledrcsources.com and www.tc.columbia.edu/lessonstudy/tools.html.

Since the teacher actually teaching the lesson already shoulders considerable responsibility, other group members may want to take responsibility for the support tasks. If your research lesson is to be observed by people beyond your immediate planning group, ask someone outside your planning group to facilitate the post-lesson discussion.

## Step 5. Discuss and Analyze the Research Lesson

In Japan, there are whole books devoted to the etiquette of lesson study discussion. No wonder! A US teacher described feeling "naked" as she taught a research lesson in front of colleagues. A good agenda for the discussion can go a long way toward easing this sense of vulnerability and making the discussion safe and productive.

Discussions of research lessons vary greatly in Japan, depending on the size and familiarity of the group attending the lesson, and whether it includes educators beyond those who planned the lesson. Several features of the lesson discussion guidelines portrayed in Figure 19 are noteworthy:

- The teacher who actually taught the research lesson speaks first, and has the chance to point out any difficulties in the lesson before they can be pointed out by others. (And there seems to be an unwritten rule that teachers don't further criticize something that's already been identified as a problem.)

- As a rule, the lesson belongs to the whole lesson study group. It is "our" lesson, not "my" lesson, and this is reflected in everyone's speech. Group members assume responsibility for explaining the thinking and planning behind the lesson.

- The instructor or the teachers who planned the lesson should talk about why they planned it, the differences between what they planned and what actually happened, and the aspects they want observers to evaluate.[21]

- Discussion focuses on the data that were collected by the observers. Observers talk specifically about the student work and conversations they recorded; they do not speak impressionistically about the quality of the lesson.

- Free discussion time is limited; hence there is limited opportunity for "grandstanding" and digression.

Although it may seem unwise to limit free discussion, Japanese teachers often note that the end of the formal discussion is "the beginning, not the end." They assume that discussions and individual feedback on the lesson will continue informally. As one teacher said at the end of the discussion following a research lesson:

> *The research lesson is not over yet. It's not a one-time lesson; rather, it gives me a chance to continue consulting with other teachers. For example, I may say to other teachers, "I want to ask you about my last lesson you saw..." Then, the other teachers can provide me with concrete suggestions and advice because they have seen at least one lesson I con-*

*ducted. We teachers can better connect with each other in this way.[22]*

In Japan, there is a shared understanding of the nature and purpose of the discussion following a research lesson. US teachers will need to create this shared understanding. Some strategies that might be helpful are:

- Post and review the agenda and ground rules for the discussion, so that all attendees understand what kind of discussion is hoped for.

- Have a discussion chair who keeps time and facilitates, and agree in advance on how the facilitator will handle lengthy or inappropriate comments.

- Have a well-conceived plan for collecting and presenting data, so that a rich discussion is supported.

- Have one group member take notes, which will later be used by the lesson study group to think about where to go next; "we've got that in our notes" can be a great way to move the discussion on.

- Reflect in speech that the research lesson belongs to the group, not just the teacher who taught it. The instructing teacher needs to feel supported!

## Step 6. Reflect on Your Lesson Study and Plan the Next Steps

You've now completed one cycle of lesson study, from thinking about your goals to bringing them to life in an actual lesson and reflecting on what went well and what needs further work. Now's the time to think about what your group would like to do next. Would you like to make further improvements in this lesson? Would other members of your group like to try out this lesson in their own classrooms? Teachers at Paterson Public School Number Two revise and re-teach each lesson more than once, an approach highly recommended by Makoto Yoshida (Figure 10, Chapter 5). Are you happy with the goals of your lesson study effort and your group's method of operation? The following questions may help you reflect on the lesson study cycle and think about your next steps.[23]

1. What is useful or valuable about our lesson study work together?

2. Is lesson study leading us to think in new ways about our everyday practice?

3. Is lesson study helping us develop our knowledge of subject matter and of student learning and development?

4. Is our lesson study goal compelling to all of us?

5. Are we working together in a productive and supportive way?

6. Have we made progress toward our overall lesson study goal?

7. Do all members of our group feel included and valued?

8. Do non-participants in our work feel informed and invited?

It takes courage for US teachers to pioneer lesson study. Your lesson study group has taken risks to break down the walls that often isolate American teachers. Formal thanks from colleagues and administrators are in order. You have demonstrated your commitment to self-improvement in ways that may ripple through your school and beyond. Whatever else you do, be sure to congratulate and celebrate!

# Chapter 6 Notes

[1]  Jacqueline Hurd, Interview, May 10, 2001.

[2]  Class discussion comment by first-year teacher, Mills College class, January 16, 2001.

[3]  These were posted on a middle school bulletin board in Concord, California.  Unfortunately, I have not been able to ascertain their author.

[4]  Yukinobu Okada, Comments at Greenwich Japanese School Open House, November 13, 2000.

[5]  Fernandez, C., Chokshi, S., Cannon, J. , & Yoshida, M. (2002).  Learning about lesson study in the United States.  In E. Beauchamp (Ed.), *New and old voices on Japanese education*.  Armonk, N.Y.: M.E. Sharpe.

[6]  *Can you lift 100 kilograms?* is an 18-minute video of the lesson study cycle in a Japanese school, available from lessonresearch.net.

[7]  Mary Pat O'Connell, Personal communication, San Mateo Foster City Unified School District, January 16, 2002.

[8]  *Can you lift 100 kilograms?*, loc. cit.

[9]  Shinichi Togami, Lead Teacher, Inogashira Elementary School, Musashino-shi, Interview, July 2, 1996.

[10]  Yoshida, M. (1999). "Lesson study: A case study of a Japanese approach to improving instruction through school-based teacher development." Doctoral dissertation, University of Chicago, 427.

[11]  Lewis, C. & Tsuchida, I. (1998, Winter).  A lesson is like a swiftly flowing river:  Research lessons and the improvement of Japanese education. *American Educator*, 51.

[12]  Sir Isaac Newton in a letter to his colleague Robert Hooke, dated February 5, 1676.

[13]  Heather Crawford, Interview, March 14, 2001.

[14]  Ibid.

[15]  Kazuyoshi Morita,  Interview, July 3, 1996.

[16]  Nakamura, T. Zadankai: Shougakkou ni okeru jugyou kenkyuu no arikata wo kangaeru. (Panel discussion: Considering the nature of lesson study in elementary schools) in Ishikawa, K., Hayakawa, K., Fujinaka, T., Nakamura, T., Moriya, I., & Takii, A. (2001). *Nihon Suugaku Kyouiku Gakkai Zasshi (Journal of Japan Society of Mathematical Education)*, 84:4,18.

[17]  See Appendix 2, levers lesson plan, 1.

[18]  Nakamura, T., loc. cit.

[19]  Ishikawa, K. Zadankai: Shougakkou ni okeru jugyou kenkyuu no arikata wo kangaeru. (Panel Discussion: Considering the nature of lesson study in elementary schools) in Ishikawa, K., Hayakawa, K., Fujinaka, T., Nakamura, T., Moriya, I., & Takii, A. (2001). *Nihon Suugaku Kyouiku Gakkai Zasshi, (Journal of Japan Society of Mathematical Education)*, 84:4, 18.

[20]  Fernandez, Chokshi, Cannon, & Yoshida, (2001), loc. cit.

[21]  Takahashi, A. Kenkyuu jugyou wo dou norikiru ka (How to navigate a research lesson).  In Shin Sansuu Kyouiku Kenkyuukai (New Mathematics Education Research Group) (Ed.) *Jugyou kenkyuu no susumekata, fukamekata, norikirikata (How to encourage, deepen, and navigate lesson study)*.  Tokyo:  Tooyoukan Publishing Company, 58, translation by Etsuko Tobari.

[22]  Lewis & Tsuchida (1998, Winter), op. cit., 16.

[23] These reflection questions are influenced by the work of Clea Fernandez and Makoto Yoshida with the Lesson Study Research Group (www.tc.columbia.edu/lessonstudy).

# 7

# Supports for Lesson Study

**The real voyage of discovery lies not in seeking new landscapes but in having new eyes.**

**- Marcel Proust (1871-1922)**

Japan and the US have different educational systems and cultures. What supports will be needed if lesson study is to succeed in the US? This chapter explores one institutional support — a shared, frugal curriculum — and five values that support lesson study: self-criticism; openness to outsiders; embrace of mistakes; willingness to borrow; and honest, respectful feedback.

## Have a Shared, Frugal Curriculum

Lesson study focuses on *how* to teach, not just *what* to teach. When US teachers spend time combing through massive textbooks to figure out what subject matter should be taught or to locate lessons relevant to standards, these may be important activities. But they are only first steps toward lesson study, which focuses on how students will learn the material.

Common standards and a shared curriculum are likely to ease lesson study for US teachers. In turn, lesson study can provide an excellent vehicle for bringing to life curriculum and standards. However, US textbooks, as well as some standards, may provide impossibly vague or voluminous expectations about what needs to be learned.

The Japanese elementary curriculum is frugal. It specifies several broad, overarching goals for each subject area as well as a *small* number of specific topics and content goals within each subject area.[1] For example, the overarching goals for elementary science include love of nature, ability to solve problems, active efforts to understand natural phenomena, and scientific habits of mind.[2] Japanese fifth-graders study just seven topics over their 95 periods of science.[3] For one of the seven topics, pendulums and weight, the *Course of Study* specifies just the following content:

> *Students come to hold ideas about the laws of motion by using weights to investigate motion, and by changing the amount of weight, speed of movement, etc. [They understand that]*
>
> a. *The cycle time for a weight suspended on a line is not changed by the amount of weight suspended, but is changed by the length of the line.*

> *b. The function of a weight moving another object varies*
> *with the weight's heaviness and speed of motion.[4]*

The other topics are similarly frugal in the amount of content to be learned. With an average of 13 to 14 lessons devoted to a single topic like the study of pendulums and weight, it makes sense for teachers to spend time crafting ways to interest students and to promote deep, lasting understanding of the content. In contrast, some American teachers may feel pressured to "cover" a topic like pendulums and weight in just one period, packing it into a lecture or presentation. If so, there may be little point in studying students' thinking and engagement.

But US teachers who are faced with voluminous standards or textbooks needn't throw in the towel. By identifying the topics that are persistently difficult for students, by finding the topics that are prominent across standards and textbooks, and by sharing thoughts with colleagues about the content that is truly central to a subject, teachers can zero in on a good focus for lesson study.

## Remain Self-Critical

A funny thing happened to me during the months in 1993 when I sat in Japanese elementary classrooms. Woven throughout all of classroom life is *hansei* ("han-say"), reflection that is often practiced at the end of individual lessons, the school day, the week, the term, and so forth. In *hansei*, students ask themselves questions like "Did I try my very hardest?" "Did I remember all my needed materials for school this week?" "Did I do anything kind for others?" and "What areas do I need to improve?" Figure 20 shows some student work focused on *hansei*. As students and teachers earnestly reflected on their behavior, it was contagious. I started asking myself whether I had done my very best at my research, and what I needed to improve. The habit of self-critical reflection is a key support for lesson study (and for Japanese education more broadly).[5]

The spirit of *hansei* — open, honest reflection focused on improvement of one's shortcomings — is a central value of lesson study. In Japan, research lessons by renowned elementary teachers attract thousands of teachers, but even these renowned teachers do not regard their lessons as models or perfect specimens. Instead, they focus on the aspects of classroom instruction they are trying to improve. The logic seems to be that their strengths are already strong; so why should they focus on them? When self-critique and improvement effort are esteemed by others, it creates a very favorable climate for instructional improvement.

A study of school improvement in six US districts underlines the importance of offering one's own practice for improvement. One quality that distinguished the successful schools was the willingness of program leaders and administrators to share their own instruction and invite critique. Their willingness to open up their own teaching to others and to share their own shortcomings sent a powerful message about the value placed on self-improvement.[6] Likewise, by volunteering to teach research lessons, lesson study pioneers can break the ice for others.

**Figure 20**
**Examples of Student Reflections** *(Hansei)*

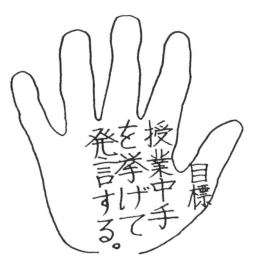

My goal: To raise my hand and speak during class. (Student name)

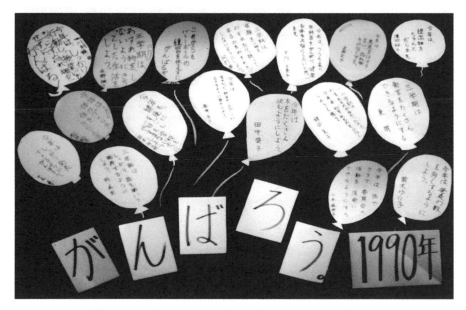

**Goals written and signed by each student:**
"I'm going to read lots of books this year."
"I'm going to be able to volunteer my ideas often next quarter."
"I'm going to keep going to volleyball practices and trying my best."
"I'm going to do my club, committee, and chore activities energetically."
"I'm going to try to be healthy this year too, and not miss any school."

*Supports for Lesson Study*

Principal Lynn Liptak's willingness to teach a mathematics class and to put her own practice on the line for scrutiny epitomizes this view of leaders as learners. The growth of lesson study at Paterson School Number Two, considered by some an unlikely site, probably rested in part on Principal Liptak's "learnership." She has written:

> Traditionally, the principal's questions, confusion, struggles, and ambiguity are kept out of view. In practical inquiry, the questions, errors, curiosity, and vulnerability of the principal become part of the constructive process.... I honestly believe that the main job of the principal is to learn and to help other people to learn...the principal is not the instructional leader but a lead learner.[7]

When asked what attitudes were essential to the success of lesson study, a San Mateo teacher answered:

> That you can always get better at teaching. That you're never at the end of the road.... If you came into [lesson study] and you were [acting] like "I'm the hottest thing out there and I've got all these great ideas and I'll share them with you guys"... you're not going to get anything out of it.[8]

However wonderful a research lesson is, something can always be improved, if not in the lesson itself, then in some other aspect of student development or classroom climate. An essential value underlying lesson study is to see this gap not as a threat or embarrassment, but as a wonderful challenge and a spur to one's fullest development as an educator.

## Remain Open to Outsiders

Educational leaders face a dilemma. If an innovation like lesson study is mandatory, reluctant participants can ruin it for everyone else; but if lesson study is voluntary, a lesson study "clique" can develop, polarizing a staff. Sharing one's teaching is an intensely personal, risky activity. Hence members of a lesson study group are likely to develop strong bonds over time. Care must be taken to keep a group open to new members and to fully integrate any newcomer who joins. Although it may be tempting to keep one's lesson study group cozy and unchanged, it is worth remembering the benefits for students when an instructional philosophy begins to appear consistently across classrooms.

## Embrace Mistakes

Real change takes time and effort and involves making errors along the way. Perfect lessons are as rare as perfect human beings. Lesson study is not a quick fix, but a slow, steady means for teachers to improve instruction, and to build a school and district culture focused on inquiry and improvement. Perhaps the most surprising threat to lesson study in the US is the enthusiasts who expect to see perfect research lessons and become discouraged or dismissive when they don't.

The yardstick for measuring the worth of a research lesson is how much teachers learned as they planned, conducted, and discussed the lesson, and whether this learning will improve their future instruction. One can learn a great deal from imperfect lessons, perhaps even more than from highly polished lessons. All nine lessons taught by Japanese and US teachers during the summer 2001 lesson study workshop in San Mateo-Foster City encountered challenges, as teachers ran out of time, discovered students' prior knowledge was not as expected, and discovered problems with the manipulatives or questions they had chosen. Yet teachers reported that they learned an enormous amount about teaching and about mathematics during the workshop, learning that probably stemmed from seeing the problems their lessons encountered in actual practice, and from the chance to discuss these problems with thoughtful colleagues. One lesson instructor, Judith Hitchings, began the discussion of a research lesson she taught by sharing a quote from the wall of San Jose's Tech Museum of Innovation: "If everything you try works, you are not trying hard enough."[8] Could there be a better motto for lesson study?

## Don't Worship Originality

What's important about a research lesson is not whether the lesson plan is original, but whether it promotes student learning. It is much better to teach a widely-available lesson that really helps students understand the nature of fractions than to teach a fully original lesson that doesn't do the job as well. It is much better to focus on a key topic in your subject area and make small improvements to the current approach than on a peripheral topic where you could do something completely original. Only if teachers build on available lessons, focusing on continuous improvement of content that is central to each discipline, will lesson study spark broad improvement.

Educator and researcher Deborah Loewenberg Ball relates the emphasis on originality of teaching to the isolation of US teachers:

> Because discussions of teaching sometimes resemble "style shows" more than they do professional interaction, teachers' development of their practice is often a highly individual and idiosyncratic matter. The common view that "each teacher has to find his or her own style" is a direct result of a discourse of practice that maintains the individualism and isolation of teaching.[9]

In contrast, the assumption supporting lesson study is that teachers cannot greatly improve children's lives *except* by working together to see that students develop the knowledge, habits of mind, and personal qualities needed to learn.

## Avoid the Twin Shoals of Happy Talk and Harping

Most of us have people in our lives, for example a spouse, relative, or close friend, whom we value precisely because we know that person will give us honest feedback, in a supportive manner, when we ask for it. With careful work, a lesson study group can become this kind of ally. The group must navigate between two dangerous extremes: happy talk (where members shy away from disagreement or criticism) and harping (where members feel and act as if preserving their own egos depends on deflating others). Ball notes the value of disagreement:

> *Masking disagreements hides individual struggles to practice wisely and so removes a good opportunity for learning. Politely refraining from critique and challenge, teachers have no forum for debating and improving their understandings.*[10]

Asked what it is like to have research lessons criticized by colleagues, Japanese teachers often reply that critical feedback is a mark of respect. As one said, "Colleagues offer criticism because they expect you can improve, and because there is something in your teaching worth improving on. What would really be scary is if they remained silent."[11]

Now that we have identified some key supports for lesson study — including a concise curriculum, self-critical stance, and willingness to learn from others and from mistakes — we move on to explore common misconceptions about lesson study.

# Chapter 7 Notes

[1] Lewis, C. (1995). *Educating hearts and minds: Reflections on Japanese preschool and elementary education*. New York: Cambridge University Press.

[2] Monbusho (Ministry of Education) (1999). *Shougakkou gakushuu shidou youryou kaisetsu: Rikahen (Explanation of the Course of Study for Elementary Schools: Science)*. Tokyo: Toyokan Publishing Co., 9.

[3] Periods last 45 minutes, although teachers may combine them to teach science in double periods.

[4] Monbusho (1999), op. cit., 51.

[5] See Lewis, C. (1995), loc. cit.

[6] Solomon, D., Battistich, D., Watson, M., Schaps, E. & Lewis, C. (2000). A six-district study of educational change: Direct and mediated effects of the child development project. *Social Psychology of Education*, 4, 3-51; Lewis, C. Journey of change. Unpublished manuscript.

[7] This account of Paterson School Number Two is based heavily on Wang-Iverson, P., Liptak, L., and Jackson, W. (2000). "Journey beyond TIMSS: Rethinking professional development." Paper presented at International Conference on Mathematics Education, Hangzhou, China; Research for Better Schools (2000, Fall). Against the odds, America's lesson study laboratory emerges, *RBS Currents*, 4:1, 8-10 (http://www.rbs.org); and Fernandez, C., Chokshi, S., Cannon, J., & Yoshida, M. (2001). Learning about lesson study in the United States. In E. Beauchamp (Ed.), *New and old voices on Japanese education*. Armonk, N.Y.: M.E. Sharpe.

[8] Gordon Moore, on the wall of the Tech Museum of Innovation in San Jose, California, undated.

[9] Ball, D.L. (1996, March). Teacher learning and the mathematics reforms: What we think we know and what we need to learn. *Phi Delta Kappan*, 505.

[10] Ibid.

[11] Group interview with visiting teacher-researchers from Morioka, at Tsukuba Elementary School, November 19, 1996.

# 8

# Misconceptions about Lesson Study

**Lesson study is not so much about lesson planning as it is about research and watching children learn.**

- US Teacher[1]

It's natural to reshape a new idea, such as lesson study, into a familiar mold, such as lesson planning. Over time, educators will no doubt want to adapt lesson study to US circumstances. But at the outset, it's important to understand how lesson study differs from practices that are common in the US. This chapter explores six misconceptions about lesson study: lesson study is simply lesson planning; lessons should be written from scratch; lessons are rigidly "scripted;" the point of lesson study is to write and spread "perfect" lessons; the research lesson is a demonstration lesson; and lesson study is basic research.

## Misconception 1: Lesson Study Is Lesson Planning

Lesson planning is just a small part of lesson study, a larger process that includes formulating long-term goals, studying student responses to an actual lesson, and revising the approach to instruction. Even the lesson planning itself differs from that familiar to most Americans. In lesson study, teachers formulate the questions and activities that will move students from their current understanding to the desired understanding of the subject matter. They anticipate student thinking in response to these activities, and they compare this imagined lesson "drama" to what actually occurs when the lesson is taught. The actual student reactions are used to revamp the lesson plan, and the instructional approach more generally. Lesson study focuses not just on a single lesson, but on the larger unit and instructional vision of which it is a part. The word lesson (*jugyou*) also means "lessons" and "instruction."

## Misconception 2: Lesson Study Means Writing Lessons from Scratch

Japanese teachers rarely develop whole new lessons from scratch during lesson study. Rather, they study state-of-the-art approaches and refine these or adapt them to their setting. The focus is on improvement, not creation. Before embarking on a research lesson, it makes sense to search out good available lessons on the topic. (An outside specialist can be very useful in this.) The less writing from scratch your group does, the more you time you will have to anticipate student responses, study student work, and refine the lesson so it works for your students.

## Misconception 3: Lesson Study Means Writing a Rigid "Script"

Several US educators have told me that Japanese lessons are inappropriate for the US because they are rigidly "scripted." What does this mean? Japanese research lessons may be "scripted" in the sense that teachers carefully choose the problem or question they will use to promote student thinking. For example, the teachers studied by Makoto Yoshida selected 12 minus 7 as the best problem to introduce subtraction with regrouping, after rejecting several other problems that they thought would be too easy to solve by simple counting (e.g., 12 minus 9), or too large to visualize easily. A basic idea underlying lesson study is that the content, wording, and presentation of a problem or activity can affect student learning.

Second, Japanese teachers sometimes use the word "script" (or "drama") to describe the overall flow of the lesson: the questions the teacher will pose, the solutions and thinking that students will offer, and the experiences that will help students build their understanding. This "script" provides a way for teachers to think through the particular questions they will ask and how they can use particular student responses (or their absence) to advance the lesson.

When US teachers hear that Japanese lessons are "scripted," what may come to mind, however, is a rigid script that teachers read and follow slavishly. This is not the case. Experienced Japanese lesson study practitioners advise that, for research lessons, teachers carefully develop instructional materials and then, "...forget the teaching materials you spent time making, and instead teach students by looking at each of their faces."[2] Japanese teachers recognize that a lesson is a "swiftly flowing river" in which many decisions must be made in the moment, and that the departures made from a lesson plan often yield important insights for improving a lesson. One well-established Japanese lesson study group is called the Polar Exploration Method.[3] Educators chose this name to call attention to what they see as a similarity between teaching and polar exploration. Both demand extraordinary expertise, rigorous training, and advance planning, but at any moment carefully-laid plans may need to be abandoned when an unpredictable event, such as an Arctic storm, arises.

## Misconception 4: Lesson Study Is Writing the "Perfect" Lesson to Be Spread to Others

I am often asked whether the point of lesson study is to perfect lessons that are then somehow certified and spread. In fact, spread of lessons is largely up to individual teachers who see lessons they like and choose to adapt them for their own use. Occasionally professional associations or individual educators will publish volumes of lessons on a particular topic or theme.

Because the world is diverse and constantly changing, there is no guarantee that a particular lesson is right for all students in all schools, or that it will continue to work well with future students. Lesson study provides a means for teachers to continue to refine lessons so that they can respond effectively to the students in their class today (not yesterday). One Japanese teacher, when asked why so many schools adopted research themes related to fostering students' "initiative" and "desire to learn," answered:

> *Thirty years ago, Japanese students sat quietly and listened to everything their teachers said, and worked hard to learn whatever the teacher asked. Today's students are different. They are used to TV and video games. They don't have a long attention span. Teachers no longer have such exalted status in Japan. So we need to work hard to interest children in science and every other subject. They aren't interested just because we tell them to be. We must design lessons very well so children want to learn science and want to pursue questions themselves.[4]*

Sound familiar? Because children's lives are changing, lesson study never ends. The "lesson" of lesson study really refers to the whole of instruction. Is every student learning and growing? Are our materials, instructional techniques, and human relationships fostering our most important long-term goals for students? Even though the lesson study of subtraction with regrouping may end when a good problem is found and a good manipulative designed, other observations made during research lessons (for example, how the students treated one another or took initiative) may stimulate the lesson study in a new direction. So the goals and content of lesson study shift over time, but the lesson study itself continues.

## Misconception 5: The Research Lesson Is a Demonstration Lesson or Expert Lesson

Something striking about Japanese lesson study is the co-equal status of all participants. Roles are rotated so that all participants learn together as equals, rather than one participant acting as mentor or leader. Despite the differences in experience that may occur in a group of teachers, it is assumed that every member will have something important to contribute to lesson study, be it the fresh pair of eyes that someone new to teaching brings, or years of experience. US teachers who have viewed *Can You Lift 100 Kilograms?* often comment on the care with which the lesson instructor elicits the ideas of every other teacher in the lesson study group. All group members take responsibility for helping to develop the approach and for collection and analysis of data on the students.

Lesson study may adapt nicely to programs of mentoring, coaching, or demonstration lessons, but those situations will require some careful thought. For example, if only less experienced or less capable teachers teach research lessons, the process may be seen as reme-dial. If only expert mentors teach research lessons, the idea may take hold that they are per-fect models to imitate (or to deride, behind their backs) rather than catalysts to spark study, reflection, and improvement.

## Misconception 6: Lesson Study Is Basic Research

The term "lesson study" could equally well be translated as "lesson research" or "instruc-tional research." Japanese teachers consider lesson study to be research, and they often in-clude in the lesson study concept map a "hypothesis" about the changes in instruction that will help students develop in the desired directions. However, lesson study differs in two important ways from most US educational research (and even from some action research).

First, the primary goal of the lesson study is not to generate knowledge that others will ap-ply. It is to improve instruction for the students in one's own purview, both directly through the research lesson and indirectly through what teachers learn from the process and apply in their future teaching. Japanese educators widely share their lesson research, but there is no assumption that what works in one setting will work in another. The primary goal is always to improve instruction in one's own setting, and to document the instruction richly so that others can understand and draw on it if they wish to try something similar. Control groups, reliability checks, statistical inferences, observers "blind" to hypotheses, and other research trappings focused on generalizing one's findings to other settings do not come into play. On the other hand, the primary importance of helping the particular students in this setting means that the classroom is scoured for evidence of every kind regarding how the students responded to the lesson and what they learned from it.

Second, lesson study examines an active improvement effort, not just any idea or question. The question "Why do some children actively participate in science problem solving, while others don't?" might guide action research or inquiry in the US. For lesson study, such a question would need to be reframed into an active intervention. For example it might be reframed as: "Does a compelling problem (such as lifting a 100-kilogram sack) encourage student participation in science problem solving?" The lesson redesign would then be tried and student participation studied. The point of lesson study is not to isolate particular vari-ables and study their effects individually, but to practice all the qualities thought to comprise good teaching, and to do this not just for the research lesson but every day, so colleagues can see the cumulative effects of these practices in one's classroom and school. I was puzzled when the teachers featured in the videotape *Secret of Trapezes* described their pendulums research lesson as "practice" for a large public research lesson on levers to be held in the fall. How could a lesson on pendulums be practice for a lesson on levers? Teachers ex-plained that both lessons would provide visiting teachers with a window on the school's phi-losophy of science education and on students' development of scientific curiosity, skills, and habits of mind, as well as their grasp of a particular scientific content.

In a traditional research model, research is applied to practice. In lesson study, practice is research (see Figure 3 in Chapter 2). A physician may diagnose a milk allergy by asking a patient to eliminate milk products and studying what happens. Lesson study's method is similar. For example, teachers notice a problem such as low motivation to study science, make thoughtful changes in their teaching approach, and then carefully observe whether these changes help. Both medical practice and teaching are clinical sciences, primarily concerned with improving clients' well-being and only secondarily concerned with generating knowledge to be applied elsewhere.

Just as "mistakes are a natural part of learning,"[5] misconceptions about lesson study are a natural product of efforts to understand lesson study and bring it to life in the US. They provide welcome evidence that US educators are actively making sense of lesson study.

What circumstances will enable US educators to work through misconceptions and build a robust version of lesson study in the US? That is the issue addressed by the final chapter.

# Chapter 8 Notes

[1]  A US teacher's reflections on a lesson study workshop, San Mateo, CA, August 9, 2001.

[2]Takahashi, A. Kenkyuu jugyou wo dou norikiru ka (How to navigate a research lesson).  In Shin Sansuu Kyouiku Kenkyuukai (New Mathematics Education Research Group) (Ed.), *Jugyou kenkyu no susumekata, fukumekata, norikirikata (How to encourage, deepen, and navigate lesson study).* Tokyo: Tooyoukan Publishing Company,  58, translation by Etsuko Tobari.

[3]  http://www.edu.ipa.go.jp/mirrors/rika/Kyokuchi/.

[4]  Japanese elementary teacher, Ochanomizu Attached Elementary School, June 11, 1997.

[5]  Takahashi, A. (August 10, 2001) to fifth grade students in the video *To Open A Cube.*  Available from lessonresearch.net.

# 9

# Next Steps

**Traveler, there are no roads. The road is created as we walk it together.**

**- Antonio Machado[1]**

**Few things are impossible to diligence and skill.**

**- Samuel Johnson[2]**

With its call for "something like lesson study to be tested in the United States," *The Teaching Gap* sparked considerable interest in lesson study.[3] But can lesson study be made to work in the United States? What will determine whether lesson study becomes as useful to US educators as to their Japanese colleagues, or whether it becomes one more educational fad to be tried and discarded? This chapter explores the conditions that will be needed for lesson study to thrive and be effective in the US.

Since *The Teaching Gap* was published in 1999, lesson study has been the focus of numerous state and national conferences in the US, has attracted more than 1,000 educators to various lesson study open houses, and has received attention in both professional and mass publications.[4] But the history of education in the US is filled with innovations that were initially greeted with enthusiasm and later pronounced ineffective. Lesson study could easily follow the path of so many other once-promising innovations: a rush to implementation; attention to superficial features of the innovation rather than its underlying principles; and a leap to conclude that the (poorly implemented) innovation is ineffective. As a veteran US educator succinctly put it, "We don't try anything long enough."[5]

Lesson study is a simple idea. But it is not a simple process. The goal-setting, data collection, lesson discussion, collaborative planning, and re-teaching that occur during lesson study are distinctively new to most US teachers. These processes are not yet tailored to US settings, but must be adapted through steady, thoughtful work. Although this may seem like a tall order, educators like Deborah Loewenberg Ball have suggested that all significant reform should probably be viewed not as "implementation of programs" but as "adaptation and generation of new knowledge."[6] To do this hard work of adaptation, US teachers will need to have or to create conditions such as time, a focused curriculum, and support for trial-and-error learning, with all its attendant messiness and disappointment. This is particularly true at this early stage.

Many reforms fail, as noted above, when their superficial features are implemented, but their underlying essence lost. For example, we can imagine in the case of lesson study that districts might "implement" lesson study by having teachers collaboratively produce and distribute lesson plans (one obvious feature of lesson study). However, it may turn out that the lesson plans themselves play an insignificant role in instructional improvement; it is the experience of collaborative goal-setting, planning, observation, and lesson discussion that contributes to professional growth. Innovations don't come with their features neatly labeled "superficial" and "essential"; so lesson study pioneers must identify which features are essential through careful, continuous evaluation of their work, by asking whether their lesson study effort is providing the essential experiences identified in Chapter 4, and whether it is beginning to build the benefits described in Chapter 2.

If lesson study is to grow, it will be because small groups of US teachers thoughtfully adapt this approach to US settings and share their work with others. These small groups of lesson study pioneers will succeed only if they find the approach genuinely useful; only if it helps them understand students, learning, and subject matter; and only if it helps them teach in ways that are more effective.

At the start, lesson study in the US is unlikely to provide the range of benefits that it does in Japan, where there is a well-developed network of lesson study experiences and publications. But it must provide some benefits if US educators are to be motivated to do the hard work of bringing lesson study to life in the US. US teachers quoted throughout this volume mention the professional satisfactions of connecting one's daily practice to long-term goals and the intellectual satisfaction of solving challenging problems related to student learning. Efficacy in the classroom is likely to be another important motivator for US teachers. Paterson School Number Two teacher Nick Timpone reflects:

> *Lesson study has made me much more aware of the need to engage the students in each and every lesson. How can I pique their interest? How can I vary my style of teaching? Lesson study has made me a more reflective and patient teacher. I teach with less brute force.[7]*

Lesson study may also provide a welcome force of coherence for teachers, supporting them as they make sense of reforms and translate them into classroom practice. As pictured in Figure 21, US teachers in many settings face myriad, sometimes contradictory reforms, often without adequate time or an established process to make sense of these reforms. Although the Japanese teachers pictured in Figure 22 also face tremendous pressure to change (rapid change seems to be a fact of postindustrial society), they face fewer, more coherent demands, and they face them together. Lesson study provides a systematic venue for Japanese teachers to discuss, try out, and rethink Japan's reform goals such as enabling students to "take initiative as learners" and "learn with desire." US teachers do not need one more thing on their already overfilled plates. For lesson study to succeed, US teachers must find that it is not "one more thing," but, instead, a process that helps to make sense of the barrage of current reforms and to bring them to life (or weed them out).

One cannot overestimate the importance of the lesson study pioneers who are beginning to emerge across the United States. If lesson study thrives in the US, it will be because these teachers, and others who join them, do the sustained, challenging, thoughtful work of adapting lesson study to US circumstances and because they persist in making this approach genuinely useful to US teachers who wish to improve instruction. It will be because they are willing to share their learning with others, as did the teachers at Paterson School Number Two and San Mateo-Foster City School District who courageously opened up their research lessons to educators from across the US, and whose eager mathematics students remind us all of the importance of this work. Bill Jackson, one of the originators of the School Two lesson study effort, writes:

> *I feel that the biggest mistake we can make when pitching lesson study to US teachers is to tell them that it is easy and painless. It is hard and possibly painful and they should prepare for it. The rewards, however, are fantastic. Real, concrete, observable improvement occurs in teaching.*[8]

Our lesson study pioneers will need to be brave enough to challenge the norms of privacy and isolation that pervade many schools and to take on work that is both intellectually and interpersonally demanding. Are you ready?

## Figure 21
## US Teacher Faces Pressure to Change

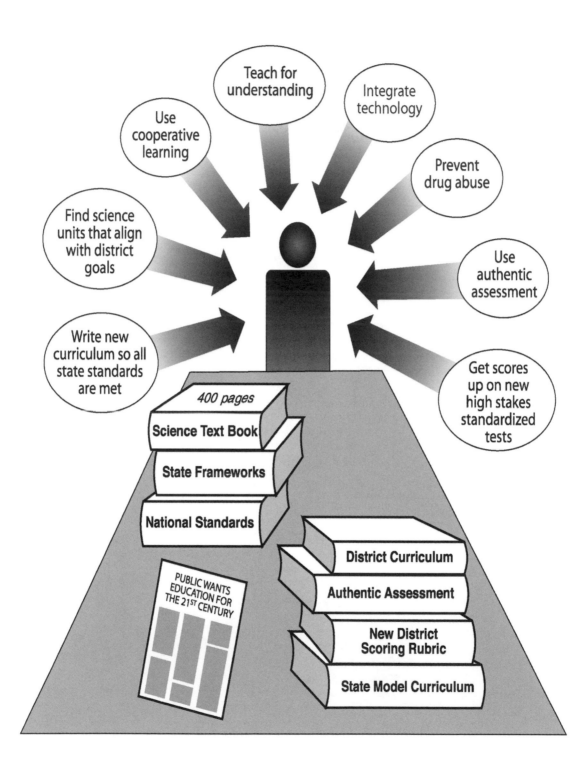

**Figure 22**
**Japanese Teachers Face Pressure to Change**

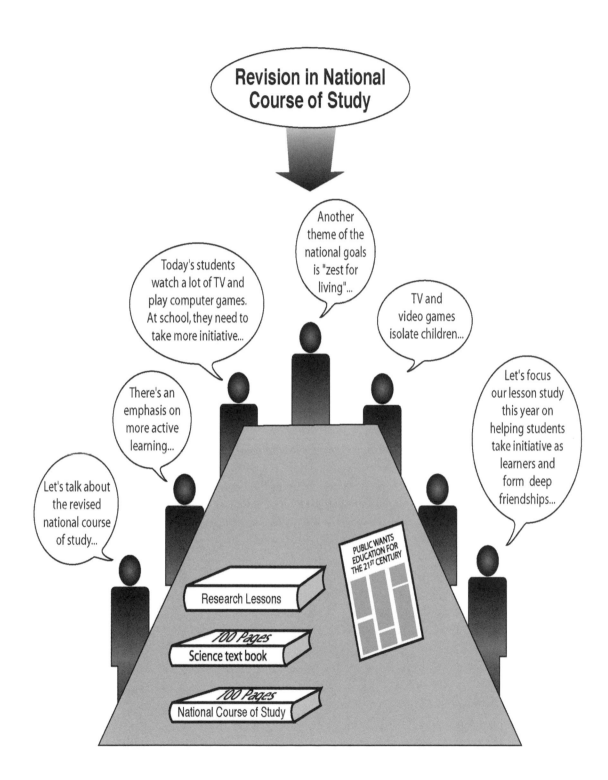

# Chapter 9 Notes

[1] From "Proverbios y Cantares, XXIX" by Antonio Machado, http://www.geocities.com/Athens/Delphi/5205/Proverbios_y_cantares.htm; translator not given.

[2] Johnson, S. (1759/1990). Rasselas and other tales. In G. Kolb (Ed.), *The Yale edition of the works of Samuel Johnson.* New Haven: Yale University Press, 56.

[3] Stigler, J.W. & Hiebert, J. (1999). *The teaching gap.* New York: Free Press, 131.

[4] For example, conferences were held by the International Congress of Teachers of Mathematics, California State Subject Matter Projects, Association of Mathematics Teachers of New Jersey, and the Greenwich Japanese School. Published references include: Coeyman, M. (2000, May 23). US school, Japanese methods. *Christian Science Monitor*; Council for Basic Education (2000). The eye of the storm: Improving teaching practices to achieve higher standards. *Briefing book, Wingspread Conference*, September 24-27, Racine Wisconsin; Germain-McCarthy, Y. (2001). *Bringing the NCTM standards to life: Exemplary practices for middle schools.* Larchmont, N.Y.: Eye on Education; and Stepanek, J. (2001, Spring). A new view of professional development. *Northwest Teacher*, 2:2, 2-5.

[5] Norma Griffin, Educator, Comments following research lesson at Pine Hollow Middle School, Concord, CA, March 4, 2002.

[6] Ball, D.L., (1996, March). Teacher learning and the mathematics reforms: What we think we know and what we need to learn. *Phi Delta Kappan*, 506.

[7] Nick Timpone, Teacher, Paterson Public School Number Two, Questionnaire Response, January, 2001.

[8] Bill Jackson, Teacher, Paterson Public School Number Two, Email, January 1, 2002.

# APPENDICES

# APPENDIX 1
## Blackboard Use and Student Note-Taking:
## Arts Developed through Lesson Study
### Makoto Yoshida

Through lesson study, Japanese teachers have developed a technical vocabulary to describe teaching and learning. The word *bansho*, which might be translated as "use and organization of the blackboard," is one example. *Bansho* is made up of two words: *ban* (board) and *sho* (writing). Because it is a technical term used by teachers, it cannot be found in an ordinary Japanese dictionary.

The importance accorded *bansho* is reflected in an elementary school principal's assessment that it is one of the three main teaching skills to promote student learning. The other two are:

- *hatsumon,* teachers' questions or activities designed to provoke students' deep thinking about a problem they are solving

- *neriage,* a process that draws out student ideas and uses them to focus students on a shared and optimal solution for a problem.

In Japanese mathematics lessons, the blackboard is used not just for posing problems, presenting student work, demonstrating solutions and procedures, and recording important concepts and formulas, but also for skillfully organizing students' thoughts and thought processes, recording ideas that emerge from student discussion, and providing a summary of the lesson. In the minds of Japanese teachers, the blackboard is not simply a place to casually jot down important messages that they want the students to remember, but a tool to help organize student thinking and discussion in order to enhance the learning experience.

The importance of the blackboard is also highlighted in the word *"bansho-keikaku,"* which can be translated as "planning for use and organization of the blackboard." Blackboard planning often occurs during lesson study, and organization of the blackboard becomes a topic of analysis during the research lesson.

Japanese teachers use several principles to plan and evaluate the use of the blackboard:

- From the board, is it easy for students and teacher to follow the flow of the lesson? Is the blackboard logically and coherently organized in order to help students understand the lesson?

- Are the goal and activities of the lesson clear on the board?

- Does the board reflect students' voices, ways of thinking, and ideas?

- Does the board show how student ideas were challenged and developed through the class discussion?

- Are the materials presented on the board meaningful to students and effective for building their understanding of the lesson?

Japanese teachers sometimes comment that students' note-taking skills reflect teachers' blackboard use. Students can't be expected to take good notes in their notebooks if they are not shown a model on the board. Many teachers prepare handouts at the beginning of the school year to help students learn to take notes. In the first trimester, the handout many include the story problem and spaces to write solutions, explanations, answers, and reflections on the lesson (for example, what students discovered, understood and felt). Some teachers also ask students to write their friends' solutions and what they thought about those solutions. The students are often expected to paste their handouts in their notebooks; so the handouts are suitably small. As the year goes on, teachers gradually provide less guidance in the handouts and ask the students to produce similar notes in their own notebooks. For example, the teacher may give only a copy of the main problem, and expect students to write explanations and reflections on their own. Notebooks are often collected by the teachers so they can be evaluated and suggestions can be made for improvement.

Planning of blackboard use seems to receive little emphasis in US mathematics instruction. Blackboards in US classrooms are often covered with materials that are unrelated to the lesson, leaving very little space to write information vital to understanding the lesson. Use of multiple blackboards in different locations (front, back, side, and portable boards) may scatter information, making it difficult for the students to know where to focus and distracting from a lesson's coherence. Information may be written and erased casually. In the US, it seems blackboards are often used to demonstrate procedures, but rarely to develop a mathematical idea collectively using student ideas. When an overhead projector is used, important information that will help students to learn the material disappears as soon as the switch is turned off.

Teachers concerned about the note-taking skills of their students might want to consider the model that can be provided through careful organization of the blackboard, as well as through direct guidance and feedback on notebooks. Lesson study provides an opportunity to explore and advance the arts of blackboard use and note-taking.

# APPENDIX 2
## Plan to Guide Learning in Science
### Research Lesson from Videotape *Can You Lift 100 Kilograms?*[*]

**1. Unit: The Way Levers Work (Grade 5)**

**2. Unit Objectives**

Enable students to investigate the design and operation of levers by changing the position from which effort is exerted and the amount of effort. To learn that:

A. If you change the position of the weight, the angle of the lever changes, even though the heaviness of the weight remains the same.

B. The lever has three major points: the fulcrum, the point of effort, and the point of resistance.

C. In the operation of a lever, there is a relationship between the amount of effort exerted and position from which effort it is exerted. When the lever balances, the amount and position of effort are related to each other according to a constant principle.

**3. Connection with Research Focus**

In their daily lives, the children use tools that involve the lever principle without realizing it. Their use of these tools is thus based upon their experience. We believe that in this unit of study, by discovering the rules and laws of the lever, the children will be able to experience anew the usefulness of tools that employ the lever principle.

If the children can discover that the rules and laws of nature are at work even in the commonplace tools of everyday life, it should be a joyful experience for them.

Also, the children will be divided into groups of several children each, and there will be a number of objectives that can be achieved only through cooperating and working together. In the midst of that activity, using the ideas of their classmates for reference, we think that the students will come to be able to give expression to ideas deeper than their own (initial) thoughts, and that is why we decided upon this unit.

**4. Actual Situation of the Students**

There are a lot of children who look forward to science classes because there are so many experiments and so much hands-on work involved. The boys and girls generally get along well and work together cooperatively, but there are one or two children in the class who have difficulty entering smoothly into such groups.

Generally, the students are serious and willing to work hard at whatever they are directed to do, but they also tend to lack the desire to come up with and try out their own ideas inde-

---

[*] Lesson videotape and full lesson plan with student work at lessonresearch.net. Translation by Christopher Weinberger.

pendently. It is noticeable that some students cannot make presentations confidently, because they don't have their own firm ideas and predictions.

### 5. Steps to Accomplish Our (Research) Focus

#### (1) Strategies for the Learning Process
Based on the actual situation of the students, our plan of instruction aims to have students grapple naturally with the subject.

#### Part One of the Unit (Lessons 1-4)
In part one of the unit, we planned the following learning flow:

Through this "learning flow," children should be able to grasp the problem at hand as their own problem and take initiative in solving it.

#### Part Two of the Unit (Lessons 5-7)
In part two, by using (calibrated) laboratory levers, allow the children to discover the rules by which the level slants and balances.

#### Part Three of the Unit (Lessons 8-9)
In part three, we plan to help the children realize that there are a great many tools that employ the lever principle even in their ordinary environments, so that in their everyday lives they could put into practice what they learn. In the new *National Course of Study,* the "weight and balance" unit presently studied at the fourth grade will be combined into this unit. With that in mind, we included the balance scale in this unit. Our overall plan is to enter the unit from learning that interests and fascinates the children, then pursue the scientific facts, and then end with learning that is useful in their everyday lives. We hope to maintain the students' interest through this unit design.

#### (2) Strategies for Curriculum Materials

Because we are trying to use materials from their everyday environment to get them interested, we have chosen to use sandbags for the weights. The students already have the experience of trying to move sandbags during their physical education classes. In addition, moving the heavy sandbags is a task that will give the students a very clear sense of the problem at hand. Furthermore, using a pole in order to move these sandbags will make the students more aware that the objects around them can, with a little engineering, become useful tools.

Among the tools that employ the lever principle, we chose one (the can-crusher) that relates to the new "environmentally aware" recycling lifestyle of modern children.

### (3) Strategies for Support and Evaluation

We will prepare a worksheet for each problem in the lesson and observe the depth and flow of the students' thoughts.  Since the students who do not have the confidence to come up with and share their own ideas will be writing them down on the worksheet, they will be able to organize their own ideas and they will leave written evidence regarding the process via which their own thinking deepened through group discussion.

Also, we want to value the *tsubuyaki* (under-the-breath exclamations) of the children during the experiments.  Since this unit is supported by team teaching, we hope that we can hear more of the students' *tsubuyaki* and, by transmitting them to the other students, help each of the individual students to deepen their own thoughts.

### (4) Strategies for the Learning Activities

Depending on the content of a lesson, there are many different ways to group children. We decided that for this unit we want to group together children who came up with similar first ideas for solving the problem.  This will make the discussions which take place within each group particularly important; by relating to other students who have similar ideas, the students should thus be able both to deepen and modify their own thoughts.

Also, feelings of competition among groups may lead students to aim at higher goals and work all the harder at devising their own group's experimental methods.

### 6.  Plan of Instruction (Unit Plan:  Nine 45-minute lessons)

| Lesson | Learning Activity | Methods; Points to Notice |
|---|---|---|
| | **Part 1:  Let's Try Moving Heavy Objects** | |
| 1. | What should be done in order to lift a heavy object off the ground?<br>• How heavy is it?<br>• Which tools should be used?<br>• Which way should it be done? | Support/Evaluation:<br>Worksheet<br>Curriculum Materials:<br>Sandbags |
| 2 and 3.<br>(The Research Lesson) | Let's really try moving it.<br>• Will it lift?<br>• How does it feel?<br>• Isn't there an easier way to lift it?<br>• Let's try the pole.<br>• Let's try changing the position we lift from.<br>• Let's change the place it's supported.<br>• Let's see how the weight feels. | Student Activities:<br>Groups of children with the same ideas<br>Curriculum Materials:<br>Sandbags, poles, and other equipment based on the children's ideas<br>Support/Evaluation:<br>Worksheet<br>Children's utterances |
| 4. | Let's find the rules for moving heavy objects.<br>• Let's think about the relation between the fulcrum and the point of effort.<br>• Let's think about the power put into the point of effort. | |

| | Part 2: What Kind of Rules Govern the Balance of the Lever? | |
|---|---|---|
| 5. | Let's discover the rules of the lever when it is tilted.<br>• Let's think about which way it tilts when we change the distance between the fulcrum and the point of resistance and fulcrum and point of effort, and when we vary the weight of the object to be lifted. | Support/Evaluation:<br>Worksheet<br>Children's utterances |
| 6. | Let's discover the rules of the lever when it is balanced.<br>• Let's try to balance the lever by changing the distance between fulcrum and point of resistance and between fulcrum and point of effort, and by varying the weight of the object to be lifted. | Support/Evaluation:<br>Worksheet<br>Children's utterances |
| 7. | Let's put together the rules of the lever when balanced and when tilted.<br>• Let's put together our understanding: the lever tilts or balances according to the value on the left and right arms on:<br>"Weight of the Object" x<br>"Distance from the Fulcrum" | |
| | Part Three: Let's Search for Tools that Use the Lever Principle | |
| 8. | Let's try using some tools that employ the principle of the lever when it is balanced.<br>• Let's try using a balance scale. | Learning Process:<br>The Balance Scale |
| 9. | Let's try using some tools that employ the principle of the lever when it is tilted.<br>• Let's try a can-crusher.<br>• Let's try a bottle-cap remover.<br>• Let's try prying nails.<br>• Let's try pliers.<br>• Let's try scissors.<br>• Let's try tweezers. | Curriculum Materials:<br>The Can Crusher |

## 7. Today's Lesson

### (1) The Aims of This Lesson

1. For students to actively consider how to use objects easily after finding that it's really hard to lift a heavy object by hand.

2. For students to deepen their own thinking by expressing their ideas to others.

3. For students to pay attention to safety and to cooperate with their friends while conducting experiments.

## (2) The Development of This Lesson

| Teacher 1 Activity | Learning Activity | Means and Points to Notice | Teacher 2 Activity |
|---|---|---|---|
| Whole Class Guidance: Check experimental conditions and suggestions to individual groups.<br><br>Whole-Class Guidance: Point out ways to direct their progress toward the next experiment. Give guidance and suggestions to groups working on "Floor Sandbag #1." | Try lifting the "Floor Sandbag #1":<br>• Using tools from their everyday environments.<br>• Having just one person lift it.<br>• By taking turns, make sure that every group member can lift it.<br><br>Make sure to see that it has really been lifted.<br>• The groups that lifted it try "Floor Sandbag #2."<br>• If they could not lift it, they try finding an easier way to lift "Floor Sandbag #1," and then take the challenge again.<br><br>Discover that a pole can be put to good use in order to move heavy objects. | ("Sandbag #1" weighs about 30 kg)<br><br>The groups, based on the ideas they came up with earlier, carry out their experiments.<br><br>Take turns to fill out their results on the worksheets, referring to each other's ideas.<br><br>("Floor Sandbag #2 weighs about 100 kg.)<br><br>Take turns to fill out their results on the worksheets, referring to each other's ideas. | Give guidance and help to each of the groups.<br><br>Give guidance and suggestions to groups working on "Floor Sandbag #2."<br><br>Whole-Class Guidance: Point out that weight was lifted by groups that used the pole well. |

## (3) Evaluation of This Lesson

1. After finding that it's very hard to lift a heavy object by hand, did the students actively consider how to use the pole to lift heavy objects easily?

2. Were the students able to deepen their thinking process by talking about their ideas with friends?

3. Were the students able to cooperate and attend to safety while performing the experiments?

**8. Student Groups for Experiments** (The original lesson plan provides the names of all students in each group.)

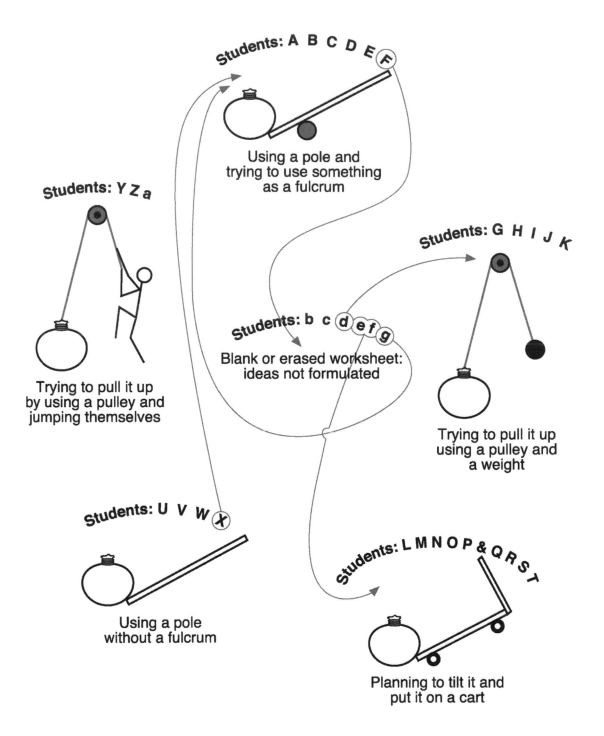

Students: A B C D E F
Using a pole and trying to use something as a fulcrum

Students: Y Z a
Trying to pull it up by using a pulley and jumping themselves

Students: b c d e f g
Blank or erased worksheet: ideas not formulated

Students: G H I J K
Trying to pull it up using a pulley and a weight

Students: U V W X
Using a pole without a fulcrum

Students: L M N O P & Q R S T
Planning to tilt it and put it on a cart

## About Team Teaching

Starting this year, eight hours of team teaching instruction per week has been introduced into our school. This year the fifth and sixth grade science classes were suited to the team teaching schedule, and we commenced this program in April. After considering the schedule and content, we scheduled team teaching out as follows:

      1st & 3rd Semesters......Sixth grade    3 hours per week x 2 classes
      2nd Semester.............Fifth grade    3 hours per week x 2 classes

The remaining two hours were used for class preparation.

When we actually carried out team teaching, we noticed several things we hadn't realized previously about team teaching. They are outlined as follows:

| Strong Points | Weak Points |
|---|---|
| • There are more opportunities to try out new kinds of experiments. <br><br> • We had three perspectives on science learning at our grade level, and we could divide the preparations among us. <br><br> • The safety of the lessons was increased. <br><br> • It was easier to hear the students' utterances, and so easier to give them support. | • Depending on the content of the lessons, such as continuing observation, or things related to weather and temperature, it is sometimes easier for the classroom teacher to conduct the lesson. <br><br> • Figuring out how to divide the time is difficult. <br><br> • Making the schedule for the use of the science classroom is also difficult. <br><br> • There is no time for planning; so it was done as follows: <br><br> Sixth Graders:    Fifth Graders: <br> Holidays         Substitute <br>                    Research Lessons |

(The instructional plan also included a copy of the student handouts for the lesson.)

# APPENDIX 3
## Plan to Guide Learning in Mathematics[1]
(Second public teaching of research lesson)


**Date:** November 27, 1998
**Students:** Grade 5, Class B (36 students)
(17 male, 19 female)
**Instructors:** Yumiko Tanaka
Michiko Honma

**Research theme: To nurture children with rich spirits who continue to learn.**

## 1. Unit: Circles and Regular Polygons

## 2. About the Unit

Circles are taught in third grade and the concept and properties of polygons are taught in an earlier unit "Congruence of triangles and quadrilaterals" in fifth grade. In this unit, we will use the hexagon-shaped toys students made for second graders to introduce the concept and properties of regular polygons and to develop strategies for drawing those shapes.

For circles, through investigation of concrete examples, and through drawing circles and measuring the circumference, students will notice that the circumference is approximately 3.14 times the diameter, and will summarize the relationship between circumference and diameter in a concrete formula.

Also, we thought the students would expand their understanding of circles through the issues that arise as they find the circumferences of many shapes, draw shapes, and find the connection between the circumference and the diameter.

Students will discover they can find the area of a circle from the radius and circumference, in an activity that uses a diagram of a partitioned circle, and draws on an area formula and the relationship between circumference and diameter students have previously learned.

In addition, we want to provide a place where students can develop their own questions as they find the area of many large shapes, can notice the relationship between radius and area, and can make shapes themselves and find their areas.

We will use team teaching to support these activities.

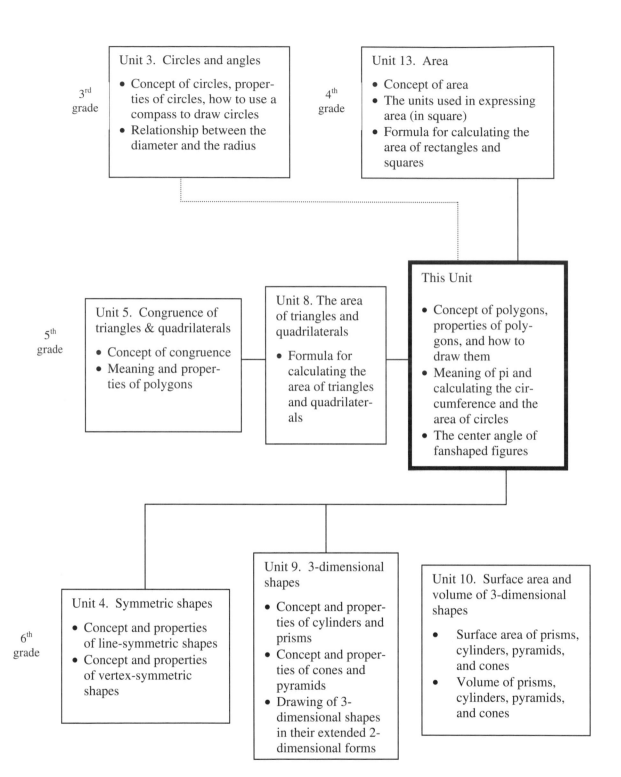

**3rd grade**

Unit 3. Circles and angles
- Concept of circles, properties of circles, how to use a compass to draw circles
- Relationship between the diameter and the radius

**4th grade**

Unit 13. Area
- Concept of area
- The units used in expressing area (in square)
- Formula for calculating the area of rectangles and squares

**5th grade**

Unit 5. Congruence of triangles & quadrilaterals
- Concept of congruence
- Meaning and properties of polygons

Unit 8. The area of triangles and quadrilaterals
- Formula for calculating the area of triangles and quadrilaterals

This Unit
- Concept of polygons, properties of polygons, and how to draw them
- Meaning of pi and calculating the circumference and the area of circles
- The center angle of fanshaped figures

**6th grade**

Unit 4. Symmetric shapes
- Concept and properties of line-symmetric shapes
- Concept and properties of vertex-symmetric shapes

Unit 9. 3-dimensional shapes
- Concept and properties of cylinders and prisms
- Concept and properties of cones and pyramids
- Drawing of 3-dimensional shapes in their extended 2-dimensional forms

Unit 10. Surface area and volume of 3-dimensional shapes
- Surface area of prisms, cylinders, pyramids, and cones
- Volume of prisms, cylinders, pyramids, and cones

### 3. Goals of the Unit

*Interest, desire to learn, and attitude:* Students will actively grapple with the task of drawing regular polygons and will take initiative to find the areas and circumferences of polygons of different sizes and shapes.

*Mathematical thinking:* Students will devise ways to draw regular polygons using a circle and will be able to consider the relationship of a circle's circumference and area to its diameter/radius.

*Expression and performance:* Students will draw regular polygons using a circle, and be able to find the circumference and area of circles.

*Knowledge and understanding:* Students will understand how to draw regular polygons using a circle. They will also understand circumference, the relationship between circumference and diameter, the meaning of pi, and the formula for calculating the area of a circle.

### 4. Current Situation of the Students

While the fifth grade students were reassigned to new homerooms, it is the third straight year this homeroom teacher has been assigned to their grade. The children in this class are cheerful and gentle, work hard to grasp various things, and are willing to learn different things. Since there are individual differences in their expression of thoughts, we are eager for students to have opportunities to express themselves.

Concerning the attitude of the students toward math, results from a survey showed that *73.5%* of the class said they "like it very much" or "like it" and *26.5%* said they "tend to dislike it" or "dislike it." That is, about one fourth of the children have some kind of resistance toward math. In addition, many children in the class indicated dislike of "using what we've learned to solve new problems" and "deriving rules and formulas from a variety of problems." However, in the unit of "division of decimals," the students enjoyed problem solving as they discovered a principle, and many students actively applied prior learning to figure out the area of new shapes in the unit "area of triangles and quadrangles." There were also more students actively trying to figure out the areas of the drawings. We hope that the students will try to think for themselves and will experience the joy of finding new things through interacting with their peers.

We also hope that students will value having their own ideas and feel pleasure in discovering new things while working with friends.

## 5. Unit Plan (12 hours of lessons)

| Hour | Objective | MAIN ACTIVITY | Nature of Instruction |
|---|---|---|---|
| 1 to 3 | To understand the concept and characteristics of polygons<br><br>To understand how to construct polygons from circles and to actually construct polygons from circles. | 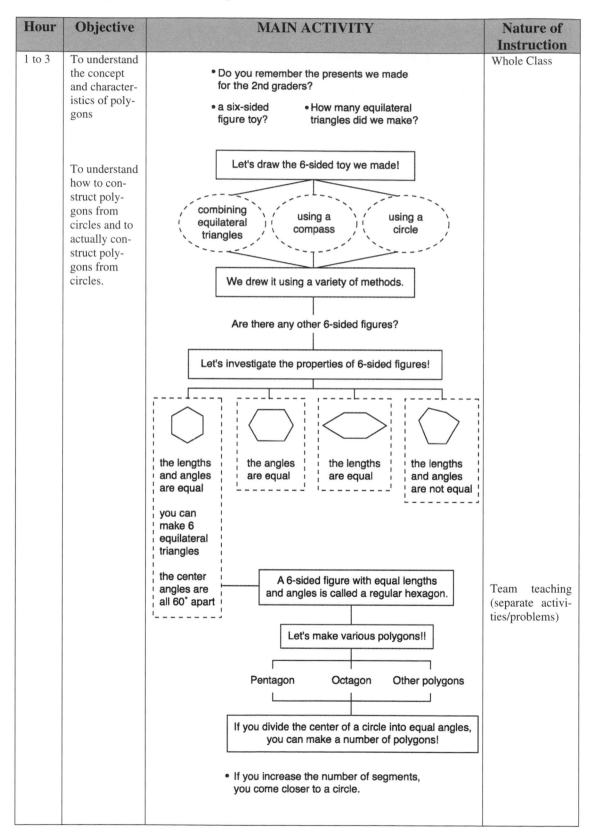 | Whole Class<br><br><br><br><br><br><br><br><br><br><br><br><br><br><br><br><br><br><br><br>Team teaching (separate activities/problems) |

| Hour | Objective | MAIN ACTIVITY | Nature of Instruction |
|------|-----------|---------------|----------------------|
| 4 to 5 | To understand the meaning of circumference and π | 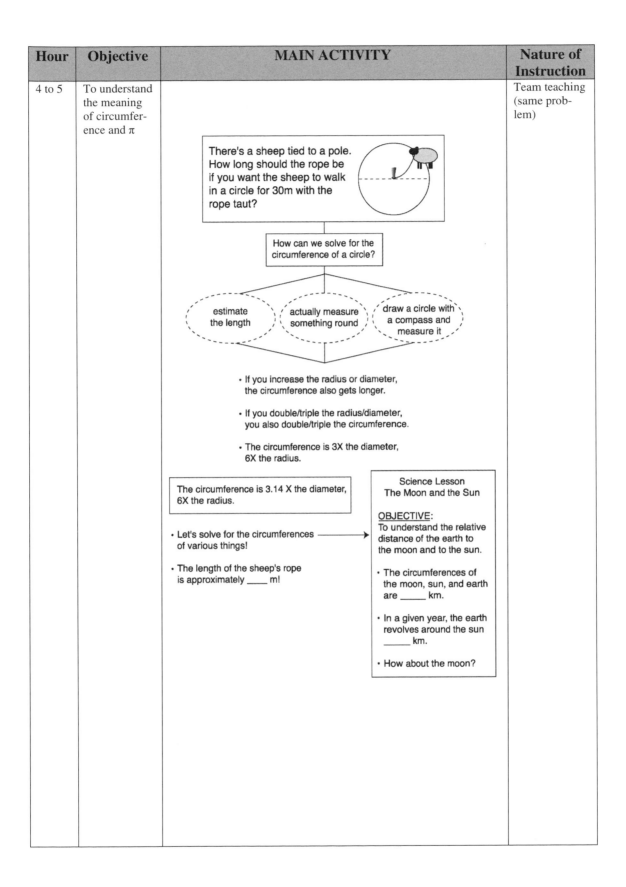 | Team teaching (same problem) |

There's a sheep tied to a pole. How long should the rope be if you want the sheep to walk in a circle for 30m with the rope taut?

How can we solve for the circumference of a circle?

estimate the length

actually measure something round

draw a circle with a compass and measure it

• If you increase the radius or diameter, the circumference also gets longer.

• If you double/triple the radius/diameter, you also double/triple the circumference.

• The circumference is 3X the diameter, 6X the radius.

The circumference is 3.14 X the diameter, 6X the radius.

• Let's solve for the circumferences of various things!

• The length of the sheep's rope is approximately _____ m!

Science Lesson
The Moon and the Sun

OBJECTIVE:
To understand the relative distance of the earth to the moon and to the sun.

• The circumferences of the moon, sun, and earth are _____ km.

• In a given year, the earth revolves around the sun _____ km.

• How about the moon?

| Hour | Objective | MAIN ACTIVITY | Nature of Instruction |
|------|-----------|---------------|------------------------|
| 6 to 7 | To find the circumferences of various circles.<br><br>To develop students' understanding of the relationship between diameter and circumference.<br><br>(Today's Lesson) | **What are the circumferences of the following figures? (Which one is the longest?)**<br><br>A      B      C<br><br>A $12 \times 3.14$<br>$=37.68$<br><br>B $12 \times 3.14 \div 2 +$<br>$6 \times 3.14 \div 2 \times 2$<br>$=37.68$<br><br>C $12 \times 3.14 \div 2 +$<br>$4 \times 3.14 \div 2 \times 3$<br>$=37.68$<br><br>It's the same!!<br><br>**Are there other figures with the same circumference?**    **Why? Is there some kind of rule?**<br><br>• There are many figures.<br>• All of them are the same length.<br>• I wonder why?<br><br>• ⌒ and ‿‿ are the same<br>• The sum of diameters are the same<br>• Are there other figures with the same relationship?<br><br>There are lots of figures with the same circumference. If the sum of the diameters is the same, the circumferences are the same | Team teaching (whole class)<br><br><br><br><br><br><br>Separate group activities |

| Hour | Objective | MAIN ACTIVITY | Nature of Instruction |
|------|-----------|---------------|----------------------|
| 8 to 9 | To under-stand how to solve for the area of a circle using estimation and trans-formation. | 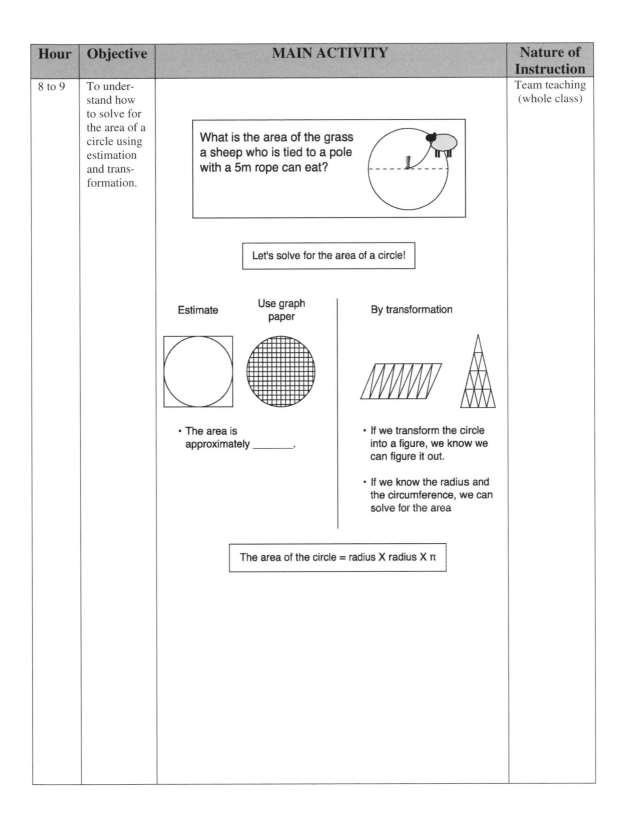 | Team teaching (whole class) |

| Hour | Objective | MAIN ACTIVITY | Nature of Instruction |
|---|---|---|---|
| 10 to 12 | To determine the area of a number of figures involving circles.<br><br>To develop students' understanding of the relationship between radius and the area of the circle. | Let's solve the area!<br><br>A    B    C    D<br><br>A 10x10x3.14 =314   B 5x5x3.14 =78.5   C 10x10x3.14÷2 +5x5x3.14 =235.5   D 10x10x3.14÷4 =78.5<br><br>If you double the radius, the area is 4X.   With a semicircle (half circle), you just divide the area of the circle by 2.<br><br>I want to investigate the relationship between radius and area.   I want to determine the areas of various figures<br><br>• If you triple the radius, you 9X the area.<br>• It's different from the relationship between diameter and circumference<br><br>• I made various figures and figured out the area!<br><br>If you double/triple the radius, you 4X/9X the area! | Team teaching (whole class)<br><br><br><br>Separate group activities |

## 6. About Instruction of This Unit

> **viewpoint 1.  develop learning activities suitable to diverse problem-solving approaches**

Teaching materials that give birth to "problem-finding"

In the beginning of November, the fifth graders made simple toys out of paper and presented them as gifts to second graders.  By recalling those hexagonal toys at the introduction of the unit, we turned students' attention to the characteristics of regular polygons and ways to draw them.

After students learned how to find circumference, they were shown three different shapes which, in contrast to their expectations, had identical circumferences.  From that experience of surprise, individual students developed a problem-consciousness and became able to notice issues like "How could it be that the shapes were different yet the circumferences the same?" and "Are there other shapes that have the same circumference?"

During the study of the area of circles, four figures were provided.  After they found the areas, the unit was designed to enable students to actively pursue important issues they noticed as they made a variety of figures and calculated their areas, and as they grasped the relationship between radius and area by comparing it with the relationship between circumference and diameter.

Through these learning approaches, we believed that children would expand their understanding of circles, find circles interesting, and also build their own capacity for continued, self-initiated learning.  Also, by presenting the shapes they discovered to the class, we expected students to deepen their motivations and reflect on their learning.

Unit structure that promotes a climate of mutual improvement

For students to continue to learn, a unit structure is needed in which students keep improving their ideas as they seek better solution methods and generalizations.  For this, solitary learning is not sufficient, and we wanted this unit to enable students to relate their ideas to one another, and recognize and build on the strengths of each other's ideas, in turn increasing the overall strength of the class.

In this unit, after discovering the methods for finding circumference and area of the circle, students formulate their own questions and think about new shapes using circles, or build on what they have previously learned.  They then can test out their ideas through interaction with peers and find new ideas.

We want students to be able to experience the sense of satisfaction and joy that comes from discovering something new, in an environment where problems are born from students' authentic experiences of surprise.  We want students to taste the pleasure of thinking together in an environment where they can make their own ideas more accurate and can find better ideas and improve them with classmates.  To create that kind of setting, it is important that each individual child has his or her own ideas.  Students understand their own strengths and those of classmates for the first time when they bring their own ideas to discussion; so we emphasized sufficient time to think individually before interaction.

We designated points that it would be good for students to discuss during small group work with friends, so that this work would connect to the larger discussion of the class. (For example: Did you discover anything? Is there anything that is always true [generalizes]?)

During whole-class discussion, the teachers should endeavor to enliven the discussions by observing students' activities and giving appropriate support (e.g., by commending previously learned knowledge; by aiding connections when students are generalizing, comparing, or deliberating; and by supporting discoveries.)

We also strategize ways to create an environment in which students could always exchange ideas easily. For example, blackboard magnets (with student names) enable students to see each other's ideas and the discussion issues, and a table workspace was created.

### Team teaching that responds to student diversity

Each individual child has different ideas when he or she approaches a problem. Especially in mathematics, there are large individual differences in depth and speed of understanding of the topic. While we want to respond to students' individual differences in style and speed of learning, and also to honor the individual differences in their ideas, at the same time we want to enable all students to experience first-hand the pleasure of reasoning. For this reason, we chose team teaching for this unit (with different forms of team teaching, such as joint activity and separate roles by task or issue). This lesson uses team teaching (separate roles, divided by issue) in the section of the lesson designed to expand students' viewpoints and thinking about circles.

In order to support students in this way and respond to each individual child, the teachers' roles have been defined in advance, and the strategies for observation and support of students agreed upon. Teachers will confer as needed during class in order to give the best support to students.

> **viewpoint 2. the evaluation and support by teachers that enable individual students to enjoy learning and improvement**

### Valuing each child's particular characteristics

Each student has special qualities. To preserve these and at the same time expand the student's potential, it is necessary to have a viewpoint from which to see that student. To accomplish this, using the class list we chose as a focus of observation each student's approach to learning, interest and desire to learn, and expressive capacity.

In addition, since more than one teacher is involved, the viewpoint for evaluation and method of guidance were agreed on before teaching, so that we could respond softly to each student. To verify and promote mathematical thinking, it is particularly important for the instructor to facilitate connections with previous learning and among viewpoints in debates/comparisons. In addition, for learning to advance, it is important for students to want to learn from each other and to notice their own and each other's strengths.

Self-evaluation strategies that enable students to experience their strengths

We tried to devise self-evaluation that would enable students to reflect on their learning and themselves. We want them both to taste their own strengths and feel a sense of efficacy of "I can do it if I try," but also a desire to go on further. We want them to reflect not just on themselves, but on the strengths of their classmates during discussion, etc., and to develop a spirit of recognizing each other and nurturing one another's growth.

With respect to evaluation, we want students to assess themselves both from the viewpoint we establish and also to choose their own viewpoint for reflection on their learning and write in a journal.

Self-Evaluation Card

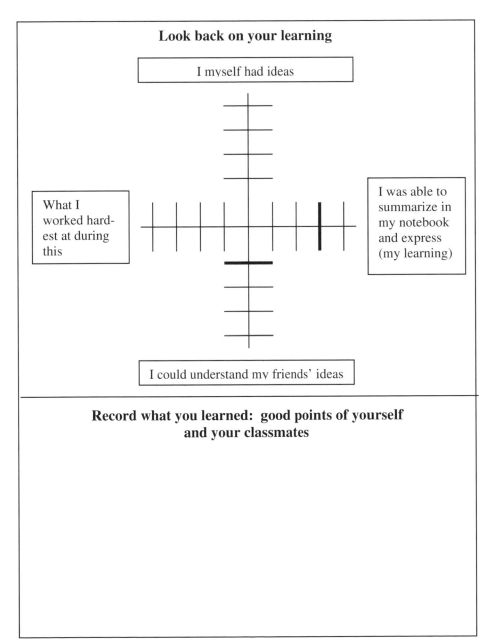

**Look back on your learning**

I myself had ideas

What I worked hardest at during this

I was able to summarize in my notebook and express (my learning)

I could understand my friends' ideas

**Record what you learned: good points of yourself and your classmates**

## 7. Lesson goals for today's lesson (7th of 12 lessons in unit).

a. While finding the circumferences of various figures made of circles, to notice the relationship between diameter and circumference, and to try to find (more) shapes with equal circumferences.

b. To try to expand one's own thinking by learning with classmates.

## 8. Lesson Plan for Today's Lesson.

| LESSON FLOW | Role of Teacher(s) |
|---|---|

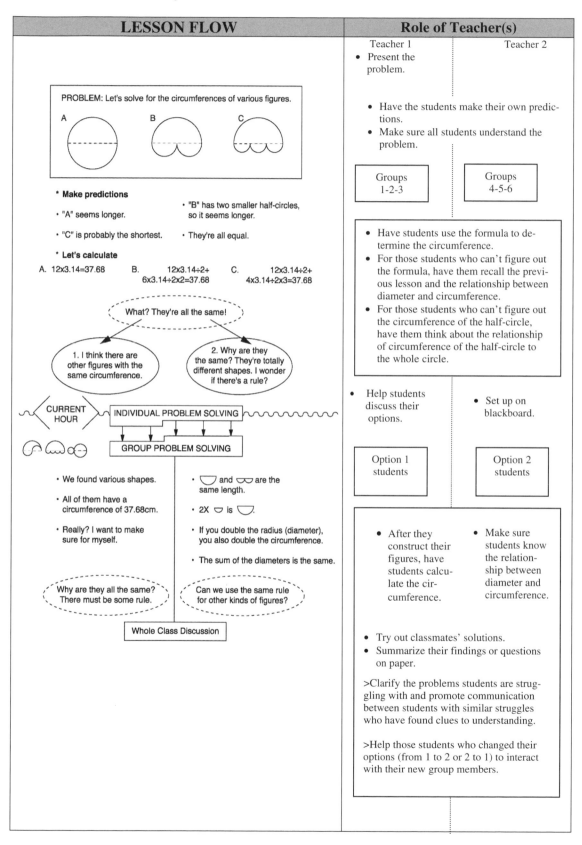

PROBLEM: Let's solve for the circumferences of various figures.

A    B    C

**\* Make predictions**

• "A" seems longer.

• "C" is probably the shortest.

• "B" has two smaller half-circles, so it seems longer.

• They're all equal.

**\* Let's calculate**

A. 12x3.14=37.68   B.   12x3.14÷2+
6x3.14÷2x2=37.68   C.   12x3.14÷2+
4x3.14÷2x3=37.68

What? They're all the same!

1. I think there are other figures with the same circumference.

2. Why are they the same? They're totally different shapes. I wonder if there's a rule?

CURRENT HOUR    INDIVIDUAL PROBLEM SOLVING

GROUP PROBLEM SOLVING

• We found various shapes.

• All of them have a circumference of 37.68cm.

• Really? I want to make sure for myself.

• ⌓ and ∽ are the same length.

• 2X ⌣ is ⌣.

• If you double the radius (diameter), you also double the circumference.

• The sum of the diameters is the same.

Why are they all the same? There must be some rule.

Can we use the same rule for other kinds of figures?

Whole Class Discussion

**Role of Teacher(s)**

Teacher 1
• Present the problem.

• Have the students make their own predictions.
• Make sure all students understand the problem.

Groups 1-2-3    Groups 4-5-6

• Have students use the formula to determine the circumference.
• For those students who can't figure out the formula, have them recall the previous lesson and the relationship between diameter and circumference.
• For those students who can't figure out the circumference of the half-circle, have them think about the relationship of circumference of the half-circle to the whole circle.

• Help students discuss their options.

Teacher 2

• Set up on blackboard.

Option 1 students    Option 2 students

• After they construct their figures, have students calculate the circumference.

• Make sure students know the relationship between diameter and circumference.

• Try out classmates' solutions.
• Summarize their findings or questions on paper.

>Clarify the problems students are struggling with and promote communication between students with similar struggles who have found clues to understanding.

>Help those students who changed their options (from 1 to 2 or 2 to 1) to interact with their new group members.

| | |
|---|---|
| **\* Discuss what they found in their small groups.**<br><br>· We discovered a number of shapes with the same circumference.<br><br>· If you double or triple the radius (diameter), you also double/triple the circumference<br><br>┌─────────────────────────────────────────┐<br>│ We found various shapes with the same circumference.<br>│ If the sum of the diameters is the same, the circumference is the same.<br>└─────────────────────────────────────────┘<br><br>**\* Reflection/summary.** | • Have students present their findings, write on the blackboard.    • Assess how students are doing.<br>                       • Help prepare students for their presentation.<br><br>Depending on the situation and students, simultaneously<br><br>• Help students realize their own and friends' efforts.<br>• Motivate students for next lesson. |

---

[1]From Dainikai *Kyouiku Jissen Happyoukai Kenkyuu Kiyou* (Second Presentation of Education Practices: Research Summary), Sapporo-shi, Hoshi-oki Higashi Public Elementary School, November 11, 1998. Thanks are expressed to the plan's authors, translators, and the researchers who made it available in the US: Michiko Honma, Shin-Ying Lee, Harold Stevenson, Hiroshi Usui, Yumiko Tanaka, and Akane Zusho.

# APPENDIX 4
## Plan to Guide Learning in Language Arts

**Date:** October 19, 1996 (Sat.) 2:00-2:45 p.m.
**Place:** Nagoya City, Kitaissha Elementary School
**Students:** Grade 3, Class 2 (37 students)
**Instructor:** Mayumi Ito

1. **Unit:  Let's Enjoy Making a Story (Special Unit)**

2. **About the Unit**

   (1) **Why we chose story making**

   Writing daily life journals is the focus of third grade composition.  It is a good opportunity for children to reflect on themselves and their lives.  Through writing daily life journals, they can consider the meaning of their actions.

   However, as children are developing and looking toward the future, they also need opportunities to forget about reality and pursue their hopes and dreams unconditionally.  When immersed in story making, children can actualize their hopes and dreams without restrictions.  Story making can free children's hearts and they can enjoy a sense of freedom different from their daily journal writing.  This unit aims at the creation of lessons in which children can express their ideas freely and enjoy story making.

   (2) **The relationship between the children's current characteristics and the unit**

   To date, the children in this class have written mostly about their experiences.  They show little resistance to writing, and many of them report that they like writing.  In addition, they like reading stories.  They enjoyed studying the textbook unit "Someko and the Ogre," guessing the feelings of Someko and the Ogre, and thinking about the continuation of the story.  I heard a child saying, "I can't wait to read the continuation of the story!"

   Since the second semester, a section called "If I were…" has been introduced to the journals which students have been writing since the first semester.  The children have enjoyed working on this section every day, and some of them wrote, "If I were a hamster, I would play hard every day on the wheel" and "If I were a bird, I would want to fly around the sky."  I look forward to reading them and respond to them every day since I can observe students' lives and wishes through the diaries.

   Looking at my students, I decided to choose story making since I wanted to let students express their feelings more freely.  It should be enjoyable and freeing for them to enter the world of their favorite tales, cartoons, and video games, and to realize their future dreams in their own stories.  I will teach this lesson "Let's enjoy making a story," hoping that my students will have a good time by writing freely.

## 3. Goals

- To enjoy making a story
- To be able to make a story based on free imagination

## 4. Unit Plan (Three 45-minute lessons)

| Learning Plan | | | Learning Goals (Points of Evaluation) | |
|---|---|---|---|---|
| Lesson | Content | Periods | Interest, Motivation | Ability to Express |
| 1 | Story making | 1 (this lesson) | Try to work on story making (behavior) Actively present own ideas (presentation) Try to listen carefully to classmates' presentations | Present title, characters, and outline (presentation) Make a story from imagination, present it, and write it (presentation, writing) |
| 2 | | 2 | Enjoy engaging in story making (attitude) | |
| 3 | Reading stories | 1 | Enjoy reading friends' stories (behavior) | Understand and enjoy the contents of friends' stories (behavior) Read friends' stories and give or write comments |

## 5. Today's Lesson

(1) Goals
- To try to engage in story making
- To engage in story making through free imagination

(2) Preparation
Story cards, hint cards, and blank paper for writing story

(3) Learning Process

| Learning Activity and Time | Points of Attention and Instruction | Evaluation Points and Method |
|---|---|---|
| 1. Understand that they will make a story. (2 min.) | T: This is a story made by A. Let's make a story like her. | |
| | Introduce a book made by a child in the class and motivate the class. | |
| 2. Discuss how to make a story. (20 min.) | T: What kind of story do you want to make? | |
| (1) Think about the content of a story and write down on a card. (5 min.) | Let children express their ideas freely and make an atmosphere in which they can enjoy making a story. | |
| | [Expected Children's Opinions] C: Happy stories C: Funny stories C: Adventures | |
| | Give a hint card to the children with no ideas. | |
| (2) Discuss what they need to decide to make a story. (5 min.) | T: What do you need to decide when you make a story? | |

| | | |
|---|---|---|
| | Give some hints so that children re-member their past experiences with stories. | |
| | [Expected Children's Opinions]<br>C: Decide the characters.<br>C: Decide what happens in the story.<br>C: Decide the title. | |
| | If no opinions are expressed, give them some suggestions. (e.g.)<br>T: What kind of title do you want to put?<br>T: Who will be in the story?<br>T: What happens in the story? | |
| (3) Decide the title and the characters and write down on a card.<br>Think about outline and present it to the class.<br>(10 min.) | T: Let's think about your own story. | |
| | Let children present their ideas freely without worrying about the order of titles and characters. Tell children that they can refer to friends' presentations if they cannot think of any ideas. | Observe student presentations and mood to see whether they were able to decide on the outline of a story.<br>A. Could make an outline and present it.<br>B. Could make an outline.<br>C. Could not make an outline. |
| | T: Let's present your story if you have some ideas. | |
| | Point out good parts of their presenta-tions, etc., in order to promote active presentation.<br>Make an atmosphere where children respect their friends' ideas by encour-aging applause to presenters. | |
| | [Expected Children's Presentation]<br>C: I will make a story of warriors' ad-venture.<br>C. I like stories with animals.<br>C: I will make a story about my friends. | |
| | Give a hint card to the children who have no ideas. [Content of hint card]<br>&bull; If X were Y…<br>&bull; X and Y<br>&bull; X which is like Y<br>&bull; X was actually Y | |
| | Give another card with actual story contents to the children who have no ideas even after looking at the hint card.<br>&bull; If Momotaro were weak…<br>&bull; If an ogre were kind…<br>&bull; A big 1$^{st}$ grader and a small 2$^{nd}$ grader<br>&bull; A strong mouse and a weak cat<br>&bull; Drawers and a cat<br>&bull; A desk and a cricket | |

| | • Scary ogre was actually kind<br>• My teacher was actually a witch | |
|---|---|---|
| | Have enough time so that many children can present their ideas on the title, the characters, and the outline.<br>Move to writing only after confirming the contents of children's presentations. | |
| 3. Write down a story and present it. (23 min.) | T: Okay, then let's make a story. | |
| (1) Write down a story. (15 min.) | Prepare two types of paper: a paper with lines and a paper with lines and a space for drawing, and let children choose one.<br>Tell children that they can also draw a picture so that they can enjoy writing a story.<br>Instruct individually and check student work.<br>Prepare extra papers for the children who need more papers. Think through a story with the children who get stuck in the middle of writing. | Observe from the nature of their writing whether students have enjoyed writing.<br>A. Able to engage enjoyably in writing<br>B. Able to engage in writing by using own card<br>C. Engaged in writing, but have not expressed ideas fully in writing. |
| (2) Present a story. | T: Please present the story you made. | |
| | Let the finished children present, providing a stimulus for others' writing and offering an opportunity to notice each other's strengths.<br>Tell children that they can present even if they did not finish yet, and create an atmosphere of eager presentation. | |

(4) Blackboard Plan

```
Let's make a story!

What kind of story do you want to make?
    •   Happy stories
    •   Funny stories
    •   Relieving stories

Things we need to decide for making a story
    •   Title
    •   Characters (Objects)
    •   Outline

(Write down the contents of children's presentations)
```

(5) Others

See the appendix for the seating chart, the hint cards, and other materials.

## Appendix

Seating Chart

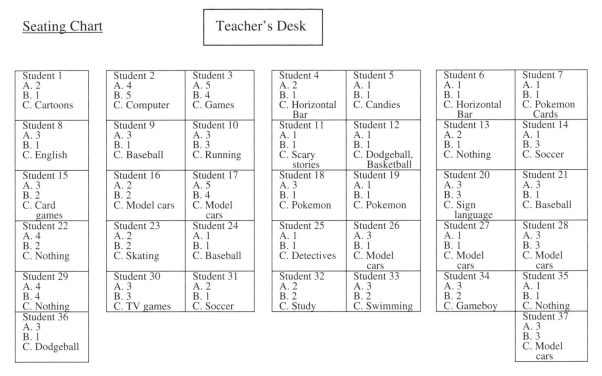

| Teacher's Desk |
| --- |

| Student 1<br>A. 2<br>B. 1<br>C. Cartoons | Student 2<br>A. 4<br>B. 5<br>C. Computer | Student 3<br>A. 5<br>B. 4<br>C. Games | | Student 4<br>A. 2<br>B. 1<br>C. Horizontal Bar | Student 5<br>A. 1<br>B. 1<br>C. Candies | | Student 6<br>A. 1<br>B. 1<br>C. Horizontal Bar | Student 7<br>A. 1<br>B. 1<br>C. Pokemon Cards |
| Student 8<br>A. 3<br>B. 1<br>C. English | Student 9<br>A. 3<br>B. 1<br>C. Baseball | Student 10<br>A. 3<br>B. 3<br>C. Running | | Student 11<br>A. 1<br>B. 1<br>C. Scary stories | Student 12<br>A. 1<br>B. 1<br>C. Dodgeball, Basketball | | Student 13<br>A. 2<br>B. 1<br>C. Nothing | Student 14<br>A. 1<br>B. 3<br>C. Soccer |
| Student 15<br>A. 3<br>B. 2<br>C. Card games | Student 16<br>A. 2<br>B. 2<br>C. Model cars | Student 17<br>A. 5<br>B. 4<br>C. Model cars | | Student 18<br>A. 3<br>B. 1<br>C. Pokemon | Student 19<br>A. 1<br>B. 1<br>C. Pokemon | | Student 20<br>A. 3<br>B. 3<br>C. Sign language | Student 21<br>A. 3<br>B. 1<br>C. Baseball |
| Student 22<br>A. 4<br>B. 2<br>C. Nothing | Student 23<br>A. 2<br>B. 2<br>C. Skating | Student 24<br>A. 1<br>B. 1<br>C. Baseball | | Student 25<br>A. 1<br>B. 1<br>C. Detectives | Student 26<br>A. 3<br>B. 1<br>C. Model cars | | Student 27<br>A. 1<br>B. 1<br>C. Model cars | Student 28<br>A. 3<br>B. 3<br>C. Model cars |
| Student 29<br>A. 4<br>B. 4<br>C. Nothing | Student 30<br>A. 3<br>B. 3<br>C. TV games | Student 31<br>A. 2<br>B. 1<br>C. Soccer | | Student 32<br>A. 2<br>B. 2<br>C. Study | Student 33<br>A. 3<br>B. 2<br>C. Swimming | | Student 34<br>A. 3<br>B. 2<br>C. Gameboy | Student 35<br>A. 1<br>B. 1<br>C. Nothing |
| Student 36<br>A. 3<br>B. 1<br>C. Dodgeball | | | | | | | Student 37<br>A. 3<br>B. 3<br>C. Model cars |

A. I like composition (1. Strongly agree, 2. Agree, 3. Neither, 4. Disagree, 5. Strongly disagree)

B. I like making a story.  (1. Strongly agree, 2. Agree, 3. Neither, 4. Disagree, 5. Strongly disagree)

C. My recent interests

Story Making Card

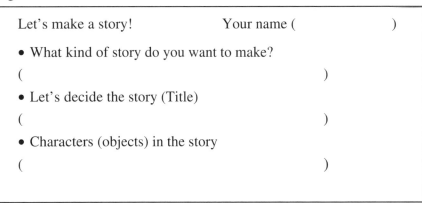

| |
| --- |
| Let's make a story!          Your name (                    )<br><br>• What kind of story do you want to make?<br>(                                          )<br><br>• Let's decide the story (Title)<br>(                                          )<br><br>• Characters (objects) in the story<br>(                                          ) |

Hint Card (1)

How should I write a story?

1. Happy story
2. Funny story
3. Touching story
4. Sad story
5. Scary story
6. Exciting story
7. Others (story of        )

Hint Card (2)

What kind of story should I make?

1. If (      ) were (      )....
2. (      ) is actually (      )
3. (      ) which is like (      )
4. (      ) and (      )

Let's try to put some words into (      ).

Hint Card (3)

What kind of story should I make?

1. If (      ) were (      )....
   - If Momotaro were weak..
   - If the Ogre were kind..

2. (      ) is actually (      ).
   - My teacher is actually a witch.
   - A scary-look ogre is actually kind.

3. (      ) which is like (      ).
   - A girl who is like a boy.
   - A school which is like an amusement park.

4. (      ) and (      )

[The plan also includes student handouts for planning and writing stories.]

# APPENDIX 5
## Plan to Guide Learning (Template)

Date:
Grade:
Subject:
School:
Instructor:
Planning Group:

1. **Unit Name**

2. **Unit Objectives**

3. **Research Theme (or "Main Aim") of Lesson Study**

4. **Current Characteristics of Students**

5. **Learning Plan for Unit:**

   - **Unit Goals or Outcomes (Connections to Standards and Prior and Subsequent Learning, if appropriate)**

   - **Sequence of Lessons in the Unit** *(The chart below may continue for several pages.)*

| Number of Lessons | Content | Points to Notice and Evaluate | Materials, Strategies |
|---|---|---|---|
| | | | |

- **Explanation of Unit "Flow" That Will Enable Students to Move from Current Understanding, Motivation, and Skills to Desired Outcomes**

**6. Plan for the Research Lesson** *(The chart below may continue for several pages.)*

| Teacher Activity | Anticipated Student Thinking and Activities | Points to Notice and Evaluate | Materials, Strategies |
|---|---|---|---|
| | | | |

a. **Aims of the Lesson**

b. **Learning Process for the Lesson (What "Drama" of Activities and Experiences Will Help Students Move from Their Initial Understanding to the Desired Aims?)**

c. **Evaluation of This Lesson (Major Points To Be Evaluated)**

d. **Copies of Lesson Materials (e.g., Blackboard Plan, Student Handouts, Visual Aids)**

7. **Background Information and Data Collection Forms for Observers (e.g., Seating Chart, Prior Student Work, Note-taking Forms, Information on Particular Students to Be Observed)**

# APPENDIX 6
# Research Map Template

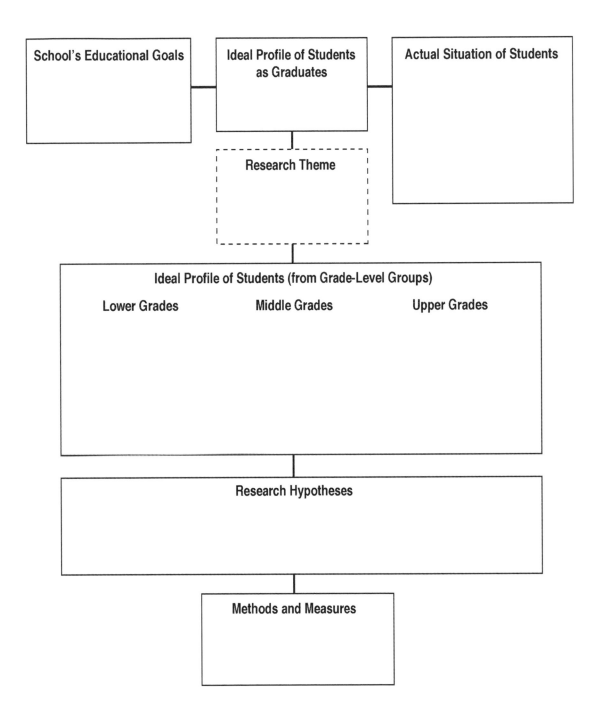

# APPENDIX 7
## Selected Resources on Lesson Study[1]
Compiled by Elizabeth King and Elizabeth Baker

**Publications:**

Boss, S. (2001, Spring). Leading from within. *Lesson study: Teachers learning together. Northwest Teacher,* 2:2.
> Focuses on how administrators can support lesson study, with a range of examples from the US. Several other articles appear in this journal issue devoted to lesson study.
> Available at: www.nwrel.org/msec/nwerc/lessonstudy.html

Lewis, C., & Tsuchida, I. (1998, Winter). A lesson is like a swiftly flowing river: Research lessons and the improvement of Japanese education. *American Educator,* 14-17 & 50-52.
> Describes the nature and purpose of the research lessons at the heart of lesson study. Quotes from Japanese teachers to highlight the impact of lesson study on teachers' professional growth, and the system supports for lesson study.
> Available at: www.lessonresearch.net

Sparks, D. (1999, November). Using lesson study to improve teaching. *Results*, National Staff Development Council.
> Succinctly makes the case for lesson study and for strong leadership support.
> Available at: www.rbs.org/lesson_study/readings_and_resources.shtml

Stigler, J., & Hiebert, J. (1999). *The teaching gap: Best ideas from the world's teachers for improving education in the classroom.* New York: Summit Books.
> Chapter 7 provides an introduction to lesson study and makes a strong case that "something like lesson study" needs to be developed in the US.

Watanabe, T. (2002). Learning from Japanese lesson study. *Educational Leadership,* 59:6, 36-39.
> A readable overview of lesson study's role in Japan and its implications for US educators.

Yoshida, M. (1999). Lesson study (*Jugyoukenkyuu*) in elementary school mathematics in Japan: A case study. Paper presented at the American Educational Research Association (1999 Annual Meeting), Montreal, Canada.
> Summarizes and highlights the work of a mathematics lesson study group at a Japanese elementary school. Rich details about the actual practice of lesson study.
> Available from: myoshida@globaledresources.com

**Websites:**

**www.globaledresources.com**
Print and video resources, professional development and consulting available to schools interested in implementing lesson study and improving mathematics. **Global Education Resources, L.L.C**

**www.lessonlab.com**
Web site supports lesson study through development and dissemination of software and related resources. **LessonLab Inc.**

**www.lessonresearch.net**
Web site features lesson study publications, news of lesson study events, weblinks, and lesson study videos to be downloaded or ordered. **Mills College US-Japan Education Project**

**www.rbs.org/lesson_study/readings_and_resources.shtml**
Web site features lesson study readings, resources, and links including the TIMSS Resource Center and links to TIMSS-related resources. **Research for Better Schools**

**www.tc.columbia.edu/lessonstudy**
Web site features many practical resources, including lesson study protocols, articles, examples of study lessons, and links to discussion forums. **Columbia Lesson Study Research Group at Teachers College, Columbia University**

### Video/Digital Media:

*Can You Lift 100 Kilograms?* (Video; 18 minutes). Highlights the lesson study cycle in a Japanese elementary school. Includes footage of lesson planning, a science research lesson, and teachers' discussion of lesson. Suitable for introduction of lesson study.
Available at: www.lessonresearch.net

**Lesson Study: An introduction** (CD-ROM). Shows steps of the lesson study cycle in mathematics, from the Japanese elementary school documented in the dissertation of Makoto Yoshida.
Available at: www.globaledresources.com

*The Secret of Trapezes* (Video; 16 minutes). Highlights of two sequential grade 5 science research lessons on pendulums described in the article "A Lesson is Like a Swiftly Flowing River," with a very brief segment of the intervening discussion.
Available from: www.lessonresearch.net

**Third International Mathematics and Science Study (TIMSS) Public Release Videotape of German, Japanese, and US Mathematics Lessons.** With an accompanying study guide, this video shows actual mathematics lessons, highlighting the instructional differences in the three countries.
Available from: National Center for Education Statistics, timss@ed.gov (telephone: 202-219-1333)

*Three Perspectives on Lesson Study* (Video, 53 minutes). Produced by the University of California Office of the President. Features presentations by Catherine Lewis ("Frequently Asked Questions About Lesson Study and Research Lessons"), Clea Fernandez ("Exploring Lesson Study in the United States"), and Jim Stigler ("What is lesson study?").
Available at: www.lessonresearch.net

---

[1] This resource lists focuses on just a few resources useful for introducing lesson study to teachers and administrators. It is drawn from a more complete Annotated Bibliography on Lesson Study compiled by Elizabeth King and Elizabeth Baker, which is available at lessonresearch.net